Arlington, Texas
May 1, 2000

Other Titles in This Series

Volume

8 **Bernd Sturmfels**
Gröbner bases and convex polytopes
1996

7 **Andy R. Magid**
Lectures on differential Galois theory
1994

6 **Dusa McDuff and Dietmar Salamon**
J-holomorphic curves and quantum cohomology
1994

5 **V. I. Arnold**
Topological invariants of plane curves and caustics
1994

4 **David M. Goldschmidt**
Group characters, symmetric functions, and the Hecke algebra
1993

3 **A. N. Varchenko and P. I. Etingof**
Why the boundary of a round drop becomes a curve of order four
1992

2 **Fritz John**
Nonlinear wave equations, formation of singularities
1990

1 **Michael H. Freedman and Feng Luo**
Selected applications of geometry to low-dimensional topology
1989

Gröbner Bases and Convex Polytopes

University
LECTURE
Series

Volume 8

Gröbner Bases and Convex Polytopes

Bernd Sturmfels

American Mathematical Society
Providence, Rhode Island

1991 *Mathematics Subject Classification*. Primary 13P10, 14M25;
Secondary 52B12, 90C10, 14Q99.

ABSTRACT. This book is about the interplay of computational commutative algebra and the theory of convex polytopes. A central theme is the study of toric ideals and their applications in integer programming. This book is aimed at graduate students in mathematics, computer science, and theoretical operations research.

Library of Congress Cataloging-in-Publication Data

Sturmfels, Bernd, 1962–
Gröbner bases and convex polytopes / Bernd Sturmfels.
p. cm. — (University lecture series, ISSN 1047-3998; v. 8)
Includes bibliographical references and index.
ISBN 0-8218-0487-1
1. Gröbner bases. 2. Convex polytopes. I. Title. II. Series: University lecture series (Providence, R.I.); 8.
QA251.3.S785 1995
512′.24—dc20
95-45780
CIP

∞ The paper used in this book is acid-free and falls within the guidelines
established to ensure permanence and durability.
Printed on recycled paper.

10 9 8 7 6 5 4 3 2 01 00 99 98 97

Contents

Introduction

Gröbner bases theory provides the foundation for many algorithms in algebraic geometry and commutative algebra, with the Buchberger algorithm acting as the engine that drives the computations. Thanks to the text books by Adams-Loustaunau (1994), Becker-Weispfenning (1993), Cox-Little-O'Shea (1992) and Eisenbud (1995), Gröbner bases are now entering the standard algebra curriculum at many universities. In view of the ubiquity of scientific problems modeled by polynomial equations, this subject is of interest not only to mathematicians, but also to an increasing number of scientists and engineers.

The interdisciplinary nature of the study of Gröbner bases is reflected by the specific applications appearing in this book. These applications lie in the domains of integer programming and computational statistics. The mathematical tools to be presented are drawn from commutative algebra, combinatorics, and polyhedral geometry.

The main thread of this book centers around a special class of ideals in a polynomial ring, namely, the class of *toric ideals*. They are characterized as those prime ideals that are generated by monomial differences, or as the defining ideals of (not necessarily normal) toric varieties. Toric ideals are intimately related to recent advances in polyhedral geometry, which grew out of the theory of $\mathcal{A}$-hypergeometric functions due to Gel'fand, Kapranov and Zelevinsky (1994). A key concept is that of a *regular triangulation*. All regular triangulations of a fixed polytope are parametrized by the vertices of the *secondary polytope*.

Both the algebra and the combinatorics appearing in this book are presented as self-contained as possible. Most of the material is accessible to first-year graduate students in mathematics. The following prerequisites will be assumed throughout:

- working knowledge of the basic facts about Gröbner bases, specifically of Chapters 1–5 and 9 of (Cox-Little-O'Shea 1992), or Chapters 1–2 of (Adams-Loustaunau 1994),
- familiarity with the terminology of polyhedral geometry and linear programming, as introduced in (Schrijver 1986) or (Ziegler 1995).

The fourteen chapters are organized as follows. In the first two chapters we present some introductory Gröbner bases material, which cannot be found in the text books. Here we consider arbitrary ideals I in a polynomial ring $S = k[x_1, \ldots, x_n]$, not just toric ideals. It is proved that I admits a *universal Gröbner basis*, that is, a finite subset which is a Gröbner basis for I with respect to all term orders simultaneously. This leads to the concept of the *Gröbner fan* and the *state polytope* of I. The state polytope is a convex polytope in $\mathbf{R}^n$ whose vertices are in bijection with the distinct initial monomial ideals of I with respect to all term orders on S. In the special case where $I = \langle f \rangle$ is a principal ideal, the state polytope coincides with the familiar Newton polytope $New(f)$. In Chapter 3 we present results and algorithms which involve the variation of term orders for a fixed ideal. In particular, it is explained how to compute the state polytope.

Chapters 4–9 deal exclusively with toric ideals. Basic algebraic features of these ideals are collected in the fourth chapter. These include degree bounds for Gröbner bases. An explicit universal Gröbner basis, the so-called *Graver basis*, is constructed.

The fifth chapter relates toric ideals to three fundamental problems associated with a (non-negative integer) linear map $\pi : \mathbf{N}^n \to \mathbf{N}^d$. Each fiber $\pi^{-1}(\mathbf{b})$, $\mathbf{b} \in \mathbf{N}^d$, consists of the lattice points in a polyhedron in $\mathbf{R}^n$. The three problems are: *enumeration* (list all points in $\pi^{-1}(\mathbf{b})$), *sampling* (pick a point in $\pi^{-1}(\mathbf{b})$ at random) and *integer programming* (find a point in $\pi^{-1}(\mathbf{b})$ which minimizes a given linear functional). The toric ideal $I_\mathcal{A}$ which is used to model these questions is the ideal of algebraic relations on the monomials with exponent vectors the columns of a matrix $\mathcal{A}$ representing π.

Chapter 6 deals with the one-dimensional case $d = 1$. In this case the Graver basis elements of $I_\mathcal{A}$ correspond to *primitive partition identities*, the variety defined by $I_\mathcal{A}$ is a monomial curve, and the associated integer program is the *knapsack problem*. Sharp degree bounds for Gröbner bases are available in this case.

Returning to our general discussion in Chapter 7, we present a geometric characterization of the universal Gröbner basis $\mathcal{U}_\mathcal{A}$, and we discuss algorithms for computing $\mathcal{U}_\mathcal{A}$. In Chapter 8 we establish a correspondence between the initial ideals of $I_\mathcal{A}$ and the regular triangulations of $\mathcal{A}$. These triangulations are parametrized by the secondary polytope $\Sigma(A)$, which is a Minkowski summand of the state polytope of $I_\mathcal{A}$. In Chapter 9 we apply our general theory to a specific family of toric ideals, namely those defined by the *second hypersimplex*.

In the next two chapters we venture into the realm of commutative algebra beyond toric ideals. Chapter 10 deals with Arnold's notion of *$\mathcal{A}$-graded algebras*. These are algebras with the simplest possible Hilbert function, namely, $1, 1, 1, 1, 1, 1, \ldots$. Their defining ideals are a natural generalization of initial ideals of toric ideals. In Chapter 11 we discuss *canonical subalgebra* bases (or *SAGBI bases*, as they were called by Robbiano & Sweedler (1990)). These bases need not be finite. But if they are, then they admit a nice geometric interpretation as degeneration of a parametrically presented variety into a toric variety.

In Chapter 12 we present advanced techniques for computing with toric ideals, and for applying them to integer programming. Chapter 13 aims to span a bridge to the theory of toric varieties as it exists in algebraic geometry, and, finally, in Chapter 14 the reader finds a collection of toric ideals and Gröbner bases which are dear to the author's heart.

At the end of each chapter there is a list of exercises and bibliographic notes. The exercises vary in difficulty: some are straightforward applications of the material presented, while others are more difficult and may lead to research projects. Many of the exercises assume a certain level of enthusiasm for performing computer experiments. The bibliographic notes are kept very brief. Their main purpose is to help the reader in locating a sample of the original or background literature. They are not intended to give a historical account or a complete bibliography of the respective subject areas.

This monograph grew out of a series of ten lectures given at the Holiday Symposium at New Mexico State University, Las Cruces, December 27–31, 1994. The material was updated and expanded considerably after the Holiday Symposium. In particular, Chapters 3, 12, 13 and 14 were added. I am grateful to numerous participants who supplied comments and helped me in locating errors in previous versions. Serkan Hosten did a particularly great job during the last round of proof-reading. I wish to thank all my co-authors listed in the bibliography for inspiring collaborations. Many of the techniques and results presented in this book I learned from and with them.

This project was supported financially by the David and Lucile Packard Foundation and the National Science Foundation through an NYI Fellowship. Most of the writing was done during my 94/95 visit at the Courant Institute of New York University. I wish to thank my hosts at NYU and at New Mexico State University for their terrific hospitality.

Special thanks go to Hyungsook and Nina, for their support, love and cheerful energy.

Bernd Sturmfels
Berkeley, August 1995

CHAPTER 1

Gröbner Basics

Let k be any field and $k[\mathbf{x}] = k[x_1, \ldots, x_n]$ the polynomial ring in n indeterminates. The monomials in $k[\mathbf{x}]$ are denoted $\mathbf{x}^{\mathbf{a}} = x_1^{a_1} x_2^{a_2} \cdots x_n^{a_n}$ and identified with lattice points $\mathbf{a} = (a_1, \ldots, a_n)$ in $\mathbf{N}^n$, where $\mathbf{N}$ stands for the non-negative integers. A total order $\prec$ on $\mathbf{N}^n$ is a *term order* if the zero vector 0 is the unique minimal element, and $\mathbf{a} \prec \mathbf{b}$ implies $\mathbf{a}+\mathbf{c} \prec \mathbf{b}+\mathbf{c}$ for all $\mathbf{a}, \mathbf{b}, \mathbf{c} \in \mathbf{N}^n$. Familiar examples of term orders are the *purely lexicographic order*, the *degree lexicographic order* and the *degree reverse lexicographic order.*

Given a term order $\prec$, every non-zero polynomial $f \in k[\mathbf{x}]$ has a unique *initial monomial*, denoted $in_\prec(f)$. If I is an ideal in $k[\mathbf{x}]$, then its *initial ideal* is the monomial ideal

$$in_\prec(I) \quad := \quad \langle \, in_\prec(f) \;:\; f \in I \, \rangle.$$

The monomials which do not lie in $in_\prec(I)$ are called *standard monomials.* A finite subset $\mathcal{G} \subset I$ is a *Gröbner basis* for I with respect to $\prec$ if $in_\prec(I)$ is generated by $\{in_\prec(g) : g \in \mathcal{G}\}$. If no monomial in this set is redundant, then the Gröbner basis $\mathcal{G}$ is *minimal.* It is called *reduced* if, for any two distinct elements $g, g' \in \mathcal{G}$, no term of g' is divisible by $in_\prec(g)$. The reduced Gröbner basis is unique for an ideal and a term order, provided one requires the coefficient of $in_\prec(g)$ in g to be 1 for each $g \in \mathcal{G}$. Starting with any set of generators for I, the *Buchberger algorithm* computes the reduced Gröbner basis $\mathcal{G}$. The *division algorithm* rewrites each polynomial f modulo I uniquely as a k-linear combination of standard monomials.

Proposition 1.1. *The (images of the) standard monomials form a k-vector space basis for the residue ring $k[\mathbf{x}]/I$.*

Clearly, there are infinitely many term orders if $n \geq 2$. However, if the ideal I is fixed, then they can be grouped into finitely many equivalence classes by the following theorem.

Theorem 1.2. *Every ideal $I \subset k[\mathbf{x}]$ has only finitely many distinct initial ideals.*

Proof: Suppose that I has an infinite set Σ_0 of distinct initial ideals. Choose a non-zero element $f_1 \in I$. Since f_1 has only finitely many terms and since each term lies in an element of Σ_0, there exists a monomial m_1 in f_1 such that the set $\Sigma_1 := \{ M \in \Sigma_0 \,:\, m_1 \in M\}$ is infinite. Since $\langle m_1 \rangle$ is strictly contained in an initial ideal of I, Proposition 1.1 tells us that the monomials outside of $\langle m_1 \rangle$ are k-linearly dependent modulo I. Hence there exists a non-zero polynomial $f_2 \in I$ none of whose terms lies in $\langle m_1 \rangle$. Since f_2 has only finitely many terms, there exists a monomial m_2 in f_2 such that the set $\Sigma_2 := \{ M \in \Sigma_1 \,:\, m_2 \in M\}$ is infinite. Since $\langle m_1, m_2 \rangle$ is strictly contained in an initial ideal of I, Proposition 1.1 tells us that the monomials outside of $\langle m_1, m_2 \rangle$ are k-linearly dependent modulo I. Hence there exists a non-zero polynomial $f_3 \in I$ none of whose terms lies in $\langle m_1, m_2\rangle$. Since f_3 has only finitely many terms, there exists a monomial m_3 in f_3 such that the set $\Sigma_3 := \{ M \in \Sigma_2 : m_3 \in M\}$ is infinite... etc...etc... Iterating this construction, we obtain an infinite strictly increasing chain of monomials ideals:

$$\langle m_1\rangle \subset \langle m_1, m_2\rangle \subset \langle m_1, m_2, m_3\rangle \subset \cdots$$

Since $k[\mathbf{x}]$ is Noetherian, this is a contradiction, and we are done. ■

Theorem 1.2 permits the following definition. A finite subset $\mathcal{U} \subset I$ is called a *universal Gröbner basis* if $\mathcal{U}$ is a Gröbner basis of I with respect to all term orders $\prec$ simultaneously.

Corollary 1.3. *Every ideal $I \subset k[\mathbf{x}]$ possesses a finite universal Gröbner basis $\mathcal{U}$.*

Proof: By Theorem 1.2, there exist only finitely many distinct reduced Gröbner bases for I. Their union $\mathcal{U}$ is finite, and it is a universal Gröbner basis for I. ■

Example 1.4. *(2×2-minors of a $2 \times m$-matrix)*
Consider a polynomial ring in $2m$ indeterminates

$$\begin{pmatrix} x_{11} & x_{12} & \cdots & x_{1m} \\ x_{21} & x_{22} & \cdots & x_{2m} \end{pmatrix}.$$

Let I be the ideal generated by the $\binom{m}{2}$ polynomials $D_{ij} := x_{1i}x_{2j} - x_{1j}x_{2i}$ for $1 \le i < j \le m$. We shall prove that this set of 2×2-minors is a universal Gröbner basis. First consider the case $m = 3$. Given any term order $\prec$, there are eight conceivable cases:

(1) $x_{11}x_{22} \succ x_{12}x_{21}$, $x_{11}x_{23} \succ x_{13}x_{21}$ and $x_{12}x_{23} \succ x_{13}x_{22}$.
(2) $x_{11}x_{22} \succ x_{12}x_{21}$, $x_{11}x_{23} \succ x_{13}x_{21}$ and $x_{12}x_{23} \prec x_{13}x_{22}$.
(3) $x_{11}x_{22} \succ x_{12}x_{21}$, $x_{11}x_{23} \prec x_{13}x_{21}$ and $x_{12}x_{23} \succ x_{13}x_{22}$.
(4) $x_{11}x_{22} \succ x_{12}x_{21}$, $x_{11}x_{23} \prec x_{13}x_{21}$ and $x_{12}x_{23} \prec x_{13}x_{22}$.
(5) $x_{11}x_{22} \prec x_{12}x_{21}$, $x_{11}x_{23} \succ x_{13}x_{21}$ and $x_{12}x_{23} \succ x_{13}x_{22}$.
(6) $x_{11}x_{22} \prec x_{12}x_{21}$, $x_{11}x_{23} \succ x_{13}x_{21}$ and $x_{12}x_{23} \prec x_{13}x_{22}$.
(7) $x_{11}x_{22} \prec x_{12}x_{21}$, $x_{11}x_{23} \prec x_{13}x_{21}$ and $x_{12}x_{23} \succ x_{13}x_{22}$.
(8) $x_{11}x_{22} \prec x_{12}x_{21}$, $x_{11}x_{23} \prec x_{13}x_{21}$ and $x_{12}x_{23} \prec x_{13}x_{22}$.

The ideal I is invariant under the action of the group S_3 by permuting columns. This induces an action on the eight cases above, with two orbits: {(1),(2),(4),(5),(7), (8)} and {(3),(6)}. Therefore it suffices to consider only the two cases (1) and (3).
Case (3): We multiply the left hand sides and right hand sides of the three given inequalities. The multiplicativity of term orders implies

$$(x_{11}x_{22}) \cdot (x_{13}x_{21}) \cdot (x_{12}x_{23}) \succ (x_{12}x_{21}) \cdot (x_{11}x_{23}) \cdot (x_{13}x_{22}).$$

Both sides are equal. This is a contradiction, hence no such term order can exist..
Case (1): We apply Buchberger's Criterion to establish the Gröbner basis property of the set $\{D_{12}, D_{13}, D_{23}\}$. The S-pair $S(D_{12}, D_{13}) = x_{23}D_{12} - x_{22}D_{13} = -x_{21}D_{23}$ reduces to zero with respect to $\{D_{23}\}$. The S-pair $S(D_{13}, D_{23}) = x_{12}D_{13} - x_{11}D_{23} = x_{13}D_{12}$ reduces to zero with respect to $\{D_{12}\}$. Finally, the minors D_{12} and D_{23} have relatively prime initial monomials, so that $S(D_{12}, D_{23})$ reduces to

zero with respect to $\{D_{12}, D_{23}\}$. This completes the proof of the universal Gröbner basis property for the three 2×2-minors of an indeterminate 2×3-matrix.

Now let $m \geq 4$ and fix a term order $\prec$. Consider any two minors D_{ij} and D_{kl}. If the set $\{i, j, k, l\}$ has four elements, then the variables in D_{ij} are disjoint from the variables in D_{kl}, hence their initial monomials are relatively prime, and $S(D_{ij}, D_{kl})$ reduces to zero with respect to $\{D_{ij}, D_{kl}\}$. If $\{i, j, k, l\}$ has three or less elements, then we may restrict to a 2×3-submatrix, and $S(D_{ij}, D_{kl})$ reduces to zero with respect to the three 2×2-minors of that submatrix by our previous discussion. We are done. ∎

Example 1.5. *(Linear Subspaces)* Let V be an $(n-d)$-dimensional vector subspace of k^n, and let I be its vanishing ideal in $k[\mathbf{x}]$. It is generated by d linear forms

$$I \quad = \quad \langle \sum_{j=1}^{n} a_{ij} x_j \ : \ i = 1, 2, \ldots, d \rangle.$$

We say that a non-zero linear form in I is a *circuit* if its set of variables is minimal with respect to inclusion. We say that a d-subset $\{j_1, \ldots, j_d\}$ of $\{1, 2, \ldots, n\}$ is a *basis* if the corresponding $d \times d$-determinant is non-zero. We abbreviate this determinant by

$$D[j_1, \ldots, j_d] \quad := \quad det \begin{pmatrix} a_{1j_1} & \cdots & a_{1j_d} \\ \vdots & \ddots & \vdots \\ a_{dj_1} & \cdots & a_{dj_d} \end{pmatrix}$$

The entries a_{ij} in these determinants are the coefficients of the linear forms defining I. It is an exercise in linear algebra to show that the circuits are precisely the non-zero linear forms

$$D[k_1, \ldots, k_{d-1}, 1] \cdot x_1 + D[k_1, \ldots, k_{d-1}, 2] \cdot x_2 + \cdots + D[k_1, \ldots, k_{d-1}, n] \cdot x_n, \quad (1.1)$$

where $1 \leq k_1 < \cdots < k_{d-1} \leq n$. Hence there are at most $\binom{n}{d-1}$ circuits (up to scaling). For the precise relationship between circuits and bases consult any book on *matroid theory.*

When applied to an ideal generated by linear forms, the Buchberger algorithm amounts to performing Gaussian elimination on the coefficient matrix. As a consequence of this, we see that every reduced Gröbner basis of I consists of precisely d circuits.

Proposition 1.6. *If I is an ideal generated by linear forms, then the set of circuits in I is a minimal universal Gröbner basis of I.*

Proof: Our discussion above shows that the set of circuits is finite and is a universal Gröbner basis. It remains to be seen that it is minimal, i.e., every circuit ℓ appears in some reduced Gröbner basis. Let X be the set of variables in ℓ. Let $\prec$ be an elimination term order such that each variable not in X comes before any variable in X, and let $\mathcal{G}$ be the reduced Gröbner basis of I with respect to $\prec$. We claim that ℓ appears in $\mathcal{G}$. If not, then there exists another linear form $\ell' \in \mathcal{G}$ with the same initial term as ℓ. By the elimination property of our term order, every variable appearing in ℓ' must lie in X. But then $\ell - \ell'$ is a non-zero linear form whose set of variables is strictly contained in X. This is a contradiction to the assumption that ℓ is a circuit. ∎

We now turn to the representation of term orders by weight vectors. Fix $\omega = (\omega_1, \ldots, \omega_n) \in \mathbf{R}^n$. For any polynomial $f = \sum c_i \cdot \mathbf{x}^{\mathbf{a}_i}$ we define the *initial form* $in_\omega(f)$ to be the sum of all terms $c_i \cdot \mathbf{x}^{\mathbf{a}_i}$ such that the inner product $\omega \cdot \mathbf{a}_i$ is maximal. For an ideal I we define the *initial ideal* to be the ideal generated by all initial forms:

$$in_\omega(I) \quad := \quad \langle\, in_\omega(f) \,:\, f \in I \,\rangle.$$

This ideal need not be a monomial ideal. However, it is whenever ω is chosen sufficiently generic. If in addition ω is non-negative, then, as we shall see, $in_\omega(I)$ is an initial monomial ideal in the earlier sense.

Example 1.7. *(Initial ideals of a principal ideal in two variables)*
Let I be the ideal generated by $f(x_1, x_2) = x_1^5x_2^2 + x_1^4x_2^4 + x_1^4 + x_1^2x_2^5 + x_1x_2^2 + x_2^6 + x_2$. If $\omega = (1,1)$ then $in_\omega(I) = \langle x_1^4x_2^4 \rangle$ is a monomial ideal. If $\omega = (1,2)$ then $in_\omega(I) = \langle x_1^4x_2^4 + x_1^2x_2^5 + x_2^6 \rangle$ is not a monomial ideal. We invite the reader to determine all initial ideals of I. How many of them are monomial ? What happens if $\omega = (0,0)$? ∎

Let $\omega \geq 0$ and let $\prec$ be an arbitrary term order. We define a new term order $\prec_\omega$ as follows: for $\mathbf{a}, \mathbf{b} \in \mathbf{N}^n$ we set

$$\mathbf{a} \prec_\omega \mathbf{b} \quad :\Longleftrightarrow \quad \omega \cdot \mathbf{a} < \omega \cdot \mathbf{b} \;\text{ or }\; (\omega \cdot \mathbf{a} = \omega \cdot \mathbf{b} \text{ and } \mathbf{a} \prec \mathbf{b}).$$

Proposition 1.8. *For every ideal $I \subset k[\mathbf{x}]$ we have $in_\prec\big(in_\omega(I)\big) = in_{\prec_\omega}(I)$.*

Proof: For every polynomial $f \in k[\mathbf{x}]$ we have $in_\prec\big(in_\omega(f)\big) = in_{\prec_\omega}(f)$. This implies that $in_\prec(in_\omega(I))$ and $in_{\prec_\omega}(I)$ contain the same monomials, and hence they are equal. ∎

Proposition 1.8 implies the following two corollaries. The first provides an algorithm for computing (Gröbner bases of) non-monomial initial ideals.

Corollary 1.9. *If $\omega \geq 0$ and $\mathcal{G}$ is a Gröbner basis of I with respect to $\prec_\omega$, then $\{\, in_\omega(g) \,:\, g \in \mathcal{G} \,\}$ is a Gröbner basis for $in_\omega(I)$ with respect to $\prec$.*

Corollary 1.10. *If $\omega \geq 0$ and $in_\omega(I)$ is a monomial ideal, then $in_\omega(I) = in_{\prec_\omega}(I)$.*

Proof: Monomial ideals remain fixed under the operation of passing to the initial ideal. ∎

Proposition 1.11. *For any term order $\prec$ and any ideal $I \subset k[\mathbf{x}]$, there exists a non-negative integer vector $\omega \in \mathbf{N}^n$ such that $in_\omega(I) = in_\prec(I)$.*

Proof: Let $\mathcal{G} = \{g_1, \ldots, g_r\}$ be the reduced Gröbner basis of I with respect to $\prec$. Write

$$g_i \quad = \quad c_{i0}\mathbf{x}^{\mathbf{a}_{i0}} + c_{i1}\mathbf{x}^{\mathbf{a}_{i1}} + \cdots + c_{ij_i}\mathbf{x}^{\mathbf{a}_{ij_i}},$$

where $in_\prec(g_i) = \mathbf{x}^{\mathbf{a}_{i0}}$. We define $C_{I,\prec}$ to be the set of all non-negative vectors $\omega \in \mathbf{R}^n_+$ such that $in_\omega(g_i) = \mathbf{x}^{\mathbf{a}_{i0}}$ for $i = 1, \ldots, r$. Equivalently,

$$C_{I,\prec} \quad = \quad \big\{\, \omega \in \mathbf{R}^n_+ \,:\, \omega \cdot (\mathbf{a}_{i0} - \mathbf{a}_{il}) > 0 \text{ for } i = 1, \ldots, r,\ l = 1, \ldots, j_i \,\big\}. \qquad (1.2)$$

We claim that $\mathcal{C}_{I,\prec}$ is non-empty. Suppose on the contrary that $\mathcal{C}_{I,\prec} = \emptyset$. By the Farkas Lemma of Linear Programming (Schrijver 1986; Section 7.8), there exist non-negative integers λ_{il}, not all zero, such that (componentwise)

$$\sum_{i=1}^{r}\sum_{l=1}^{j_i} \lambda_{il}\,(\mathbf{a}_{i0} - \mathbf{a}_{il}) \quad \leq \quad 0.$$

By the multiplicativity property of our term order, this translates into

$$\prod_{i=1}^{r}\prod_{l=1}^{j_i}(\mathbf{x}^{\mathbf{a}_{i0}})^{\lambda_{il}} \quad \preceq \quad \prod_{i=1}^{r}\prod_{l=1}^{j_i}(\mathbf{x}^{\mathbf{a}_{il}})^{\lambda_{il}}.$$

On the other hand, the requirement $in_\prec(g_i) = x^{\mathbf{a}_{i0}}$ implies $\mathbf{x}^{\mathbf{a}_{i0}} \succ \mathbf{x}^{\mathbf{a}_{il}}$, and therefore

$$\prod_{i=1}^{r}\prod_{l=1}^{j_i}(\mathbf{x}^{\mathbf{a}_{i0}})^{\lambda_{il}} \quad \succ \quad \prod_{i=1}^{r}\prod_{l=1}^{j_i}(\mathbf{x}^{\mathbf{a}_{il}})^{\lambda_{il}}.$$

This is a contradiction, and we conclude that $\mathcal{C}_{I,\prec}$ is a non-empty open convex cone.

Choose any $\omega \in \mathcal{C}_{I,\prec} \cap \mathbf{Z}^n$. We must show that $in_\omega(I) = in_\prec(I)$. The right hand ideal is generated by the monomials $in_\prec(g_i) = in_\omega(g_i) = \mathbf{x}^{\mathbf{a}_{i0}}$, and so it is clearly contained in the left hand ideal. If this containment were strict, then it would remain so after passing to the initial ideal with respect to $\prec$, which means that $in_\prec(I)$ is strictly contained in $in_{\prec_\omega}(I)$. This is impossible by Proposition 1.1. ■

If ω is any real vector such that $in_\omega(I) = in_\prec(I)$ for some term order $\prec$, then we call ω a *term order for* I. We also say that ω *represents* $\prec$ *for* I. We define the *Gröbner region* $GR(I)$ to be the set of all $\omega \in \mathbf{R}^n$ such that $in_\omega(I) = in_{\omega'}(I)$ for some $\omega' \geq 0$. Clearly, $GR(I)$ contains the non-negative orthant $\mathbf{R}^n_+$.

Example 1.7. *(continued)* The left diagram in Figure 1-1 below depicts the seven monomials appearing in $f(x_1, x_2)$. Their convex hull is a hexagon. The right diagram is the normal fan of this hexagon. The Gröbner region is the complement of the closed cone which is shaded in white:

$$GR(I) \quad = \quad \big\{(\omega_1,\omega_2) \in \mathbf{R}^2 \,:\, \omega_2 > 0 \text{ or } \omega_1 + 2\omega_2 > 0\,\big\}.$$

When dealing with problems in projective geometry, the given ideal I is generated by homogeneous polynomials. In this case the Gröbner region is all of $\mathbf{R}^n$.

Proposition 1.12. *Suppose that $I \subset k[\mathbf{x}]$ is a homogeneous ideal with respect to some positive grading $deg(x_i) = d_i > 0$. Then $GR(I) = \mathbf{R}^n$.*

Proof: Given any $\omega \in \mathbf{R}^n$, there exists $\lambda > 0$ such that $\omega' := \omega + \lambda \cdot (d_1, \ldots, d_n)$ is a positive vector. It suffices to show that $in_\omega(I) = in_{\omega'}(I)$.

If $f \in I$ is homogeneous with respect to the grading $deg(x_i) = d_i$, then we have clearly $in_\omega(f) = in_{\omega'}(f)$. Consider a non-homogeneous polynomial $f \in I$, and write $f = f_0 + f_1 + f_2 + \cdots + f_r$ for its decomposition into homogeneous components. Since I is homogeneous, we have $f_0, f_1, \ldots, f_r \in I$. The ω-initial form of f equals $in_\omega(f) = in_\omega(f_{i_1}) + \cdots + in_\omega(f_{i_s})$, for a suitable index subset $\{i_1, \ldots, i_s\}$. Each summand $in_\omega(f_{i_j})$ lies in $in_{\omega'}(I)$, and hence so does $in_\omega(f)$. This proves that $in_\omega(I) \subseteq in_{\omega'}(I)$. The proof of the reverse inclusion is analogous. ■

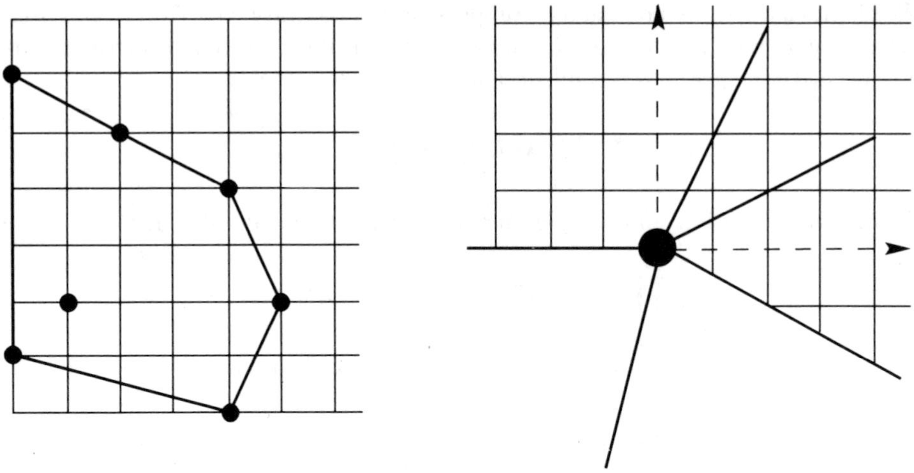

Figure 1-1. Gröbner region of a principal ideal in two variables.

We close this section with an identity which will be useful in the next chapter.

Proposition 1.13. *Let $I \subset k[\mathbf{x}]$ be an ideal, $\omega, \omega' \in \mathbf{R}^n$ and $\epsilon > 0$ sufficiently small. Then*

$$in_{\omega'}\big(in_\omega(I)\big) \quad = \quad in_{\omega+\epsilon\cdot\omega'}(I). \tag{1.3}$$

Proof: Let $\mathcal{G}$ be the reduced Gröbner basis of I with respect to the term order $\prec_{\omega+\epsilon\cdot\omega'}$. For each $g \in \mathcal{G}$ we have $in_{\omega'}(in_\omega(g)) = in_{\omega+\epsilon\cdot\omega'}(g)$. Using Corollary 1.9, this implies that the right hand side of (1.3) is contained in the left hand side, provided ϵ is sufficiently small. If this containment were proper, then it would be proper after passing to initial monomial ideals with respect to $\prec$. But this is impossible by Proposition 1.1. ∎

Exercises:

(1) Let I be the ideal generated by the nine 2×2-minors of a 3×3-matrix of indeterminates. Find a universal Gröbner basis. How many distinct initial ideals $in_\prec(I)$ are there ?

(2) Let $\mathcal{U}$ be a universal Gröbner basis for an ideal I in $k[x_1, \ldots, x_n]$. Show that, for every subset $Y \subseteq \{x_1, \ldots, x_n\}$, the elimination ideal $I \cap k[Y]$ is generated by $\mathcal{U} \cap k[Y]$. Is this property sufficient for a set $\mathcal{U}$ to be a universal Gröbner basis ?

(3) Compute all circuits in the following ideal of linear forms

$$I \quad = \quad \langle\, 2x_1 + x_2 + x_3,\;\; x_2 + 2x_4 + x_5,\;\; x_3 + x_5 + 2x_6 \,\rangle.$$

What would be a good algorithm for computing circuits in general ?

(4) Show that $in_\omega(f \cdot g) = in_\omega(f) \cdot in_\omega(g)$ for $f, g \in k[\mathbf{x}]$. Show that $in_\omega(I \cdot J) \supseteq in_\omega(I) \cdot in_\omega(J)$ for ideals I and J. Find I and J where this containment is proper.

(5) Let $I = \langle f(x_1, x_2, x_3)\rangle$ be a principal ideal for $n = 3$. Explain how to determine the Gröbner region $GR(I)$ and how to enumerate all the distinct initial ideals of I.

(6) Let I be an ideal in $k[x_1, x_2]$ generated by polynomials of degree at most d. Give an upper bound (in d) on the number of distinct initial monomial ideals of I.

Notes:

A standard reference on Gröbner bases and the Buchberger algorithm is (Buchberger 1985). The most conceptual approach to the topics in Chapters 1, 2 and 3 is the interpretation of Gröbner bases computations as a torus action on the Hilbert scheme. This geometric point of view was developed in the thesis of D. Bayer (1982). Universal Gröbner bases were introduced in (Weispfenning 1987) and (Schwartz 1988). Their existence and finiteness follows also from the general doubly-exponential degree bounds for Gröbner bases, see e.g. (Möller & Mora 1984). The more direct proof of Theorem 1.2 given here is due to A. Logar. It appeared in (Mora & Robbiano 1988). In that paper the concepts of the Gröbner fan and the Gröbner region were introduced. The representation of term orders by weight vectors was pioneered by Ostrowski (1921).

CHAPTER 2

The State Polytope

In the first half of this chapter we review some basic concepts from polyhedral geometry. In the second half we introduce the state polytope of an ideal I. It has the property that its vertices are in a natural bijection with the distinct initial ideals $in_{\prec}(I)$.

A *polyhedron* is a finite intersection of closed half-spaces in $\mathbf{R}^n$. Thus a polyhedron P can be written as $P = \{\mathbf{x} \in \mathbf{R}^n : A \cdot \mathbf{x} \leq \mathbf{b}\}$, where A is a matrix with n columns. If $\mathbf{b} = 0$ then there exist vectors $\mathbf{u}_1, \ldots, \mathbf{u}_m \in \mathbf{R}^n$ such that

$$P \quad = \quad pos(\{\mathbf{u}_1, \ldots, \mathbf{u}_m\}) \quad := \quad \{\lambda_1 \mathbf{u}_1 + \cdots + \lambda_m \mathbf{u}_m \; : \; \lambda_1, \ldots, \lambda_m \in \mathbf{R}_+\}. \quad (2.1)$$

A polyhedron of the form (2.1) is called a *(polyhedral) cone.* Here and throughout this book $\mathbf{R}_+$ denotes the non-negative reals. The *polar* of a cone C is defined as

$$C^* \quad = \quad \{\omega \in \mathbf{R}^n \; : \; \omega \cdot \mathbf{c} \leq 0 \text{ for all } \mathbf{c} \in C\}.$$

A polyhedron Q which is bounded is called a *polytope.* Every polytope Q can be written as the convex hull of a finite set of points

$$Q = conv(\{\mathbf{v}_1, \ldots, \mathbf{v}_m\}) := \{\, \sum_{i=1}^{m} \lambda_i \mathbf{v}_i \; : \; \lambda_1, \ldots, \lambda_m \in \mathbf{R}_+, \sum_{i=1}^{m} \lambda_i = 1\}. \qquad (2.2)$$

Here are two examples of 3-dimensional polytopes:

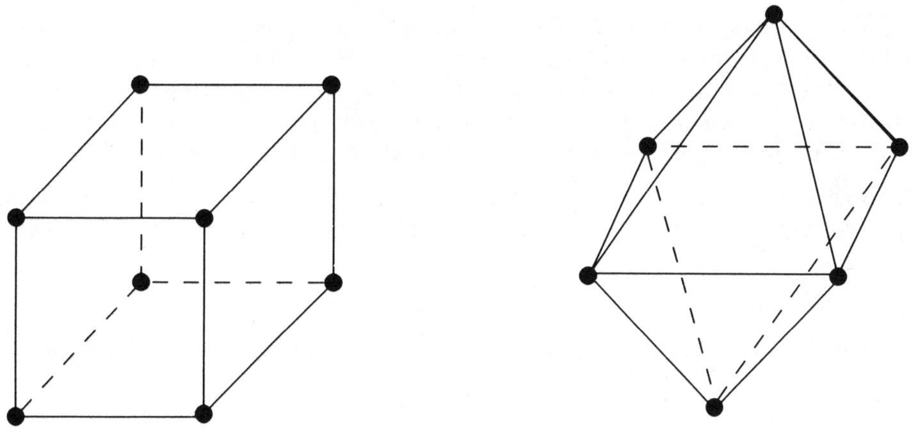

Figure 2-1. The cube and the octahedron.

Let P be any polyhedron in $\mathbf{R}^n$ and $\omega \in \mathbf{R}^n$, viewed as a linear functional. We define

$$face_\omega(P) \quad := \quad \{\mathbf{u} \in P \;:\; \omega \cdot \mathbf{u} \geq \omega \cdot \mathbf{v} \text{ for all } \mathbf{v} \in P\}.$$

Every subset F of P which has this form, that is, which maximizes some linear functional, is called a *face* of P. Note that $P = face_0(P)$ is a face of itself. For instance, each of the two polytopes in Figure 2-1 has exactly 27 faces. Their dimensions range from 0 to 3.

The relation among polyhedra "is a face of" is transitive, because of the following basic identity:

$$face_{\omega'}\big(face_\omega(P)\big) \quad = \quad face_{\omega+\epsilon\cdot\omega'}(P) \qquad \text{for } \epsilon > 0 \text{ sufficiently small.} \tag{2.3}$$

We illustrate this identity for the 3-dimensional cube. Here $face_\omega(P)$ is the upper backward edge and ω' is the vector pointing straight to the right.

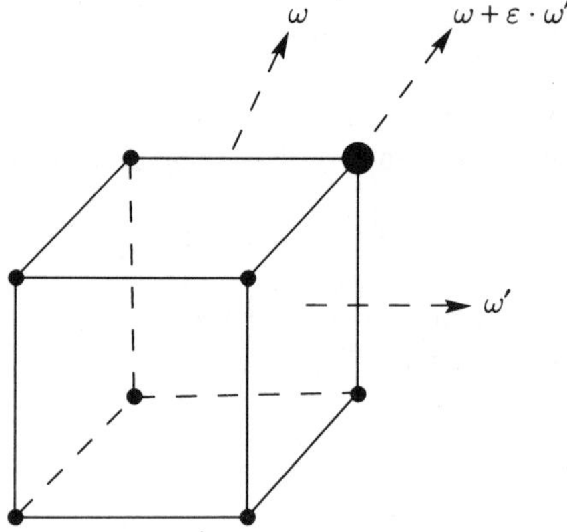

Figure 2-2. Illustration of identity (2.3).

The *dimension* of a face F of a polyhedron P is the dimension of its affine span, and its *codimension* is $codim_P(F) := dim(P) - dim(F)$. A face of codimension 1 is a *facet*. Faces of dimension 0 and 1 are called *vertices* and *edges* respectively. Every polytope is the convex hull of its vertices, and every cone is the positive hull of its edges. This makes the representations (2.2) and (2.1) unique and minimal.

Proposition 2.1. *Every polyhedron P can be written as the sum $P = Q + C$ of a polytope Q and a cone C. The cone C is unique and is called the recession cone of P.*

Proof: See Section 8.2 in (Schrijver 1986). ∎

The sum in Propositon 2.1 is defined via *Minkowski addition of polyhedra*,

$$P_1 + P_2 \quad := \quad \{\mathbf{p}_1 + \mathbf{p}_2 \;:\; \mathbf{p}_1 \in P_1,\, \mathbf{p}_2 \in P_2\}.$$

A basic fact about the Minkowski sum operation is the additivity of faces:

$$face_\omega(P_1 + P_2) \quad = \quad face_\omega(P_1) \,+\, face_\omega(P_2). \tag{2.4}$$

This implies the following remark: if $\mathbf{v}$ is a vertex of $P_1 + P_2$ then there exist unique vertices $\mathbf{p}_1$ of P_1 and $\mathbf{p}_2$ of P_2 such that $\mathbf{v} = \mathbf{p}_1 + \mathbf{p}_2$. The Minkowski sum of two polygons is computed as follows:

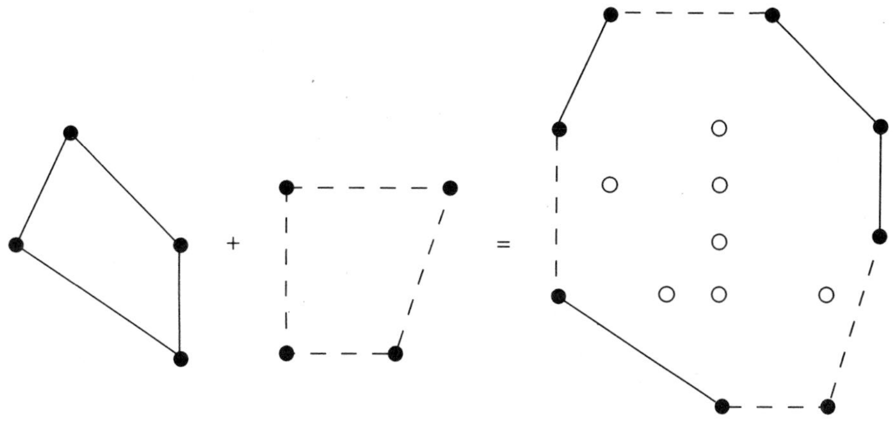

Figure 2-3. Minkowski addition of two quadrangles.

Every vertex of the Minkowski sum is a sum of vertices of the summands. Certain sums of vertices do not give vertices but interior points of the Minkowski sum. These are indicated by circles in Figure 2-3. Note also that every edge of $P_1 + P_2$ is a parallel translate of an edge of P_1 or of an edge of P_2.

A *(polyhedral) complex* Δ is a finite collection of polyhedra in $\mathbf{R}^n$ such that
(i) if $P \in \Delta$ and F is a face of P, then $F \in \Delta$;
(ii) if $P_1, P_2 \in \Delta$, then $P_1 \cap P_2$ is a face of P_1 and of P_2.
The *support* of a complex Δ is $|\Delta| := \cup_{P \in \Delta} P$. A complex Δ which consists of cones is called a *fan*. A fan Δ is *complete* if $|\Delta| = \mathbf{R}^n$.

If $P \subset \mathbf{R}^n$ is a polyhedron and F a face of P, then the *normal cone* of F at P is

$$\mathcal{N}_P(F) \quad = \quad \{\omega \in \mathbf{R}^n \,:\, face_\omega(P) = F\}.$$

Note that $dim(\mathcal{N}_P(F)) = n - dim(F)$. If F and F' are faces of P, then F' is a face of F if and only if $\mathcal{N}_P(F)$ is a face of $\mathcal{N}_P(F')$. Hence the collection of normal cones $\mathcal{N}_P(F)$, where F ranges over the faces of P, is a fan. This fan is denoted $\mathcal{N}(P)$ and called the *normal fan* of P. The support of $\mathcal{N}(P)$ equals the polar C^* of the recession cone C in the decomposition of P given in Proposition 2.1. The cone C^* has a simple interpretation in terms of linear programming: it consists of those linear functionals ω which give a bounded solution when maximized over P.

If Q is a polytope, then its recession cone is $\{0\}$, and its normal fan $\mathcal{N}(Q)$ is a complete fan. Two polytopes Q and Q' are called *strongly isomorphic* if $\mathcal{N}(Q) = \mathcal{N}(Q')$. In Figure 2-4 we depict two strongly isomorphic hexagons.

Let us now see how these concepts of polyhedral geometry are related to the primitives of computational algebra. With every polynomial $f = \sum_{i=1}^m c_i \cdot \mathbf{x}^{\mathbf{a}_i}$ in $k[\mathbf{x}]$ we associate the *Newton polytope* $New(f) := conv\{\mathbf{a}_i : i = 1, \ldots, m\}$ in $\mathbf{R}^n$. The algebraic operation of multiplication corresponds to the geometric operation of Minkowski addition:

Lemma 2.2. $New(f \cdot g) = New(f) + New(g)$.

Proof: It suffices to show that both polytopes have the same vertices. We first

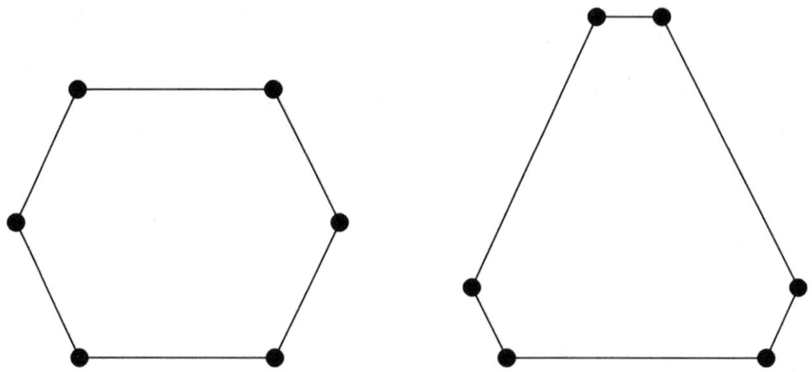

Figure 2-4. Two strongly isomorphic hexagons.

note the following general relation between faces and initial forms:

$$face_\omega(New(f)) \quad = \quad New(in_\omega(f)) \tag{2.5}$$

We next observe that Lemma 2.2 holds for monomials $f = \mathbf{x}^\mathbf{a}$ and $g = \mathbf{x}^\mathbf{b}$ because $\mathbf{x}^\mathbf{a} \cdot \mathbf{x}^\mathbf{b} = \mathbf{x}^{\mathbf{a}+\mathbf{b}}$. The following relation holds for all $\omega \in \mathbf{R}^n$ which are sufficiently generic:

$$\begin{aligned}
&face_\omega(New(f \cdot g)) =_{(2.5)} New(in_\omega(f \cdot g)) =_{\text{Exerc.(4) in Chapter1}} \\
&= New(in_\omega(f) \cdot in_\omega(g)) =_{\text{mono}} New(in_\omega(f)) + New(in_\omega(g)) =_{(2.5)} \\
&= face_\omega(New(f)) + face_\omega(New(g)) =_{(2.4)} face_\omega(New(f) + New(g)).
\end{aligned}$$

Here "sufficiently generic" means that the operator $face_\omega(\cdot)$ selects a vertex on both the left hand side and the right hand side. We conclude that the two polytopes in question have the same set of vertices and hence they are equal. ■

It is our goal to generalize the construction of the Newton polytope from polynomials to ideals. We fix an ideal $I \subset k[\mathbf{x}]$. Two weight vectors $\omega, \omega' \in \mathbf{R}^n$ are called *equivalent (with respect to I)* if and only if $in_\omega(I) = in_{\omega'}(I)$.

Proposition 2.3. *Each equivalence class of weight vectors is a relatively open convex polyhedral cone.*

Proof: Let $C[\omega]$ denote the equivalence class containing ω. We fix an arbitrary term order $\prec$ as a "tie breaker". Let $\mathcal{G}$ be the reduced Gröbner basis of I with respect to $\prec_\omega$. We claim that the following formula holds:

$$C[\omega] \quad = \quad \{\omega' \in \mathbf{R}^n \ : \ in_{\omega'}(g) = in_\omega(g) \ \text{ for all } \ g \in \mathcal{G}\}. \tag{2.6}$$

This formula implies Proposition 2.3 because it expresses $C[\omega]$ as an intersection of hyperplanes and open half-spaces. To see this, note that the right hand side of (2.6) is defined by the equations $\omega' \cdot \mathbf{a} = \omega' \cdot \mathbf{b}$ and the inequalities $\omega' \cdot \mathbf{a} > \omega' \cdot \mathbf{c}$ where $\mathbf{x}^\mathbf{a}$ and $\mathbf{x}^\mathbf{b}$ run over the terms of $in_\omega(g)$ and $\mathbf{x}^\mathbf{c}$ runs over the terms of g which do not appear in $in_\omega(g)$.

We first prove the inclusion "$\supseteq$" in (2.6). If ω' lies in the right hand side of (2.6), then $in_\omega(I) = \langle in_{\omega'}(g) : g \in \mathcal{G}\rangle \subseteq in_{\omega'}(I)$. If this inclusion of ideals were proper, then the same would be true for their initial ideals:

$$in_{\prec_\omega}(I) = in_\prec(in_\omega(I)) \quad \subset \quad in_\prec(in_{\omega'}(I)) = in_{\prec_{\omega'}}(I).$$

(Here we are using Proposition 1.8). But, in view of Proposition 1.1, it is impossible to have a proper inclusion among two initial monomial ideals of I, and therefore $in_\omega(I) = in_{\omega'}(I)$.

To prove the inclusion "$\subseteq$" in (2.6), suppose that $\omega' \in C[\omega]$. By Corollary 1.9, the set $in_\omega(\mathcal{G}) = \{in_\omega(g) : g \in \mathcal{G}\}$ is the reduced Gröbner basis of $in_\omega(I) = in_{\omega'}(I)$ with respect to $\prec$. Fix $g \in \mathcal{G}$. Then $in_{\omega'}(g)$ reduces to zero with respect to $in_\omega(\mathcal{G})$ using the term order $\prec$. This implies that the monomial $m := in_{\prec_\omega}(g)$ must appear in $in_{\omega'}(g)$, because all other monomials of g do not lie in $in_{\prec_\omega}(I) = in_\prec(in_\omega(I))$. We write $in_\omega(g) = m + h$ and $in_{\omega'}(g) = m + h'$ where h and h' are k-linear combinations of such standard monomials. After the first step of the above zero-reduction we arrive at the polynomial $h' - h$, which lies in $in_\omega(I)$. However, none of the terms appearing in $h' - h$ lies in $in_{\prec_\omega}(I) = in_\prec(in_\omega(I))$. Therefore $h' - h$ is the zero polynomial, and we conclude that $in_{\omega'}(g) = in_\omega(g)$. This completes the proof. ∎

We remark that formula (2.6) has the following geometric reformulation in terms of (normal cones of the Newton polytopes of) the reduced Gröbner basis $\mathcal{G}$:

$$C[\omega] = \mathcal{N}_Q(face_\omega(Q)), \quad \text{where } Q := New(\prod_{g\in\mathcal{G}} g) = \sum_{g\in\mathcal{G}} New(g). \tag{2.6'}$$

We define the *Gröbner fan* $GF(I)$ to be the set of closed cones $\overline{C[\omega]}$ for all $\omega \in \mathbf{R}^n$. The usage of this term is justified by the following proposition.

Proposition 2.4. *The Gröbner fan $GF(I)$ is a fan.*

Proof: Let ω' be any vector in the closure $\overline{C[\omega]}$ of an equivalence class $C[\omega]$. Then $in_\omega(I)$ is an initial ideal of $in_{\omega'}(I)$, and hence there exists a term order $\prec$ such that $in_{\prec_\omega}(I) = in_{\prec_{\omega'}}(I)$. Let $\mathcal{G}$ be the reduced Gröbner basis of I with respect to $\prec_\omega$, and let Q be the polytope defined in (2.6'). Since $\mathcal{G}$ is the reduced Gröbner basis for $\prec_{\omega'}$ as well, the equivalence classes $C[\omega]$ and $C[\omega']$ are outer normal cones of the *same* polytope:

$$C[\omega] = \mathcal{N}_Q(face_\omega(Q)) \qquad \text{and} \qquad C[\omega'] = \mathcal{N}_Q(face_{\omega'}(Q)).$$

Our hypothesis implies that the polytope $face_\omega(Q)$ is a face of the polytope $face_{\omega'}(Q)$, and therefore the closed convex cone $\overline{C[\omega']}$ is a face of the closed convex cone $\overline{C[\omega]}$,

We must show that $GF(I)$ satisfies the two axioms (i) and (ii) for being a complex. To verify axiom (i), let F be any face of the closure of an equivalence class $C[\omega]$. If ω' is any vector in the relative interior of F, then the argument in the previous paragraph shows that $F = \overline{C[\omega']}$ is a face of $\overline{C[\omega]}$. To verify axiom (ii), let $\omega, \omega' \in \mathbf{R}^n$ and consider the closed convex cone $P := \overline{C[\omega]} \cap \overline{C[\omega']}$. We have proved that, for every $\omega'' \in P$, the cone $\overline{C[\omega'']}$ is a face of $\overline{C[\omega]}$ and a face of $\overline{C[\omega']}$. Hence P is a finite union of such common faces. But an irredundant union of faces of $\overline{C[\omega]}$ can only be convex if this union is a singleton. We conclude that P is itself a common face of both $\overline{C[\omega]}$ and $\overline{C[\omega']}$. ∎

We now come to the main theorem in this chapter. From now on we shall assume that I is homogeneous with respect to some positive grading, say, $deg(x_i) = d_i > 0$.

Theorem 2.5. *Let I be a homogeneous ideal in $k[\mathbf{x}]$. There exists a polytope $State(I) \subset \mathbf{R}^n$ whose normal fan $\mathcal{N}(State(I))$ coincides with the Gröbner fan $GF(I)$.*

The polytope $State(I)$ in Theorem 2.5 will be called the *state polytope* of I. Its construction goes as follows. The degree of a monomial in the given grading is $deg(\mathbf{x}^{\mathbf{a}}) = \sum_{i=1}^n d_i a_i$. We write I_d for the vector space of homogeneous polynomials of degree d in I. If M is any monomial ideal, then $\sum M_d$ denotes the sum of all vectors $\mathbf{a} \in \mathbf{N}^n$ such that $\mathbf{x}^{\mathbf{a}}$ has degree d and lies in M. We define

$$State_d(I) \quad := \quad conv\,\Big\{ \sum in_{\prec}(I)_d \;\; : \; \prec \text{ any term order } \Big\}. \qquad (2.7)$$

Let D be the largest degree of any element in a minimal universal Gröbner basis of I. We define our polytope of interest as the Minkowski sum

$$State(I) \quad := \quad \sum_{d=1}^{D} State_d(I). \qquad (2.8)$$

The following lemma generalizes formula (2.5).

Lemma 2.6. *For all $\omega \in \mathbf{R}^n$ we have $face_\omega\big(State_d(I)\big) \; = \; State_d(\, in_\omega(I)\,)$.*

Proof: First suppose that ω is generic in $\mathbf{R}^n$, so that $in_\omega(I)$ is a monomial ideal and $face_\omega(\,\cdot\,)$ selects a vertex of the state polytope. Let $\mathbf{x}^{\mathbf{a}_1}, \ldots, \mathbf{x}^{\mathbf{a}_m}$ be all monomials of degree d, and let $r := dim(I_d) \leq m$. Let $\prec$ be any term order such that $in_\omega(I) = in_\prec(I)$. Sfter relabeling we may suppose that $\mathbf{x}^{\mathbf{a}_1}, \ldots, \mathbf{x}^{\mathbf{a}_r}$ are precisely the monomials in $in_\prec(I)_d$. Since the standard monomials form a basis modulo I, there exist polynomials

$$\mathbf{x}^{\mathbf{a}_i} \; - \sum_{j=r+1}^{m} c_{ij}\mathbf{x}^{\mathbf{a}_j} \quad \in \quad I_d \quad \text{ for } i = 1, 2, \ldots, r.$$

In each of these r equations the respectively first term is the ω-largest, i.e., $\omega \cdot \mathbf{a}_i > \omega \cdot \mathbf{a}_j$ whenever $c_{ij} \neq 0$. Recall that $State_d(in_\omega(I)) \; = \; \{\mathbf{a}_1 + \cdots + \mathbf{a}_r\}$.

The vertex $face_\omega\big(State_d(I)\big)$ is equal to $\sum in_{\prec'}(I)_d$ for some term order $\prec'$. Let $\mathbf{x}^{\mathbf{a}_{j_1}}, \ldots, \mathbf{x}^{\mathbf{a}_{j_r}}$ be the monomials in $in_{\prec'}(I)_d$. Suppose the equation claimed in Lemma 2.6 were not true. By definition of the operator $face_\omega(\,\cdot\,)$, this implies $\omega \cdot (\mathbf{a}_{j_1} + \cdots + \mathbf{a}_{j_r}) > \omega \cdot (\mathbf{a}_1 + \cdots + \mathbf{a}_r)$. By Steinitz' basis exchange lemma applied to the vector space $(k[\mathbf{x}]/I)_d$, we can pass from the sum $\mathbf{a}_1 + \cdots + \mathbf{a}_r$ to the sum $\mathbf{a}_{j_1} + \cdots + \mathbf{a}_{j_r}$ by a sequence of replacements $\mathbf{a}_i \mapsto \mathbf{a}_j$ with $c_{ij} \neq 0$ as above. We have seen in the previous paragraph that each of these replacements decreases the value of the linear functional ω. This is a contradiction. We have shown that the assertion of Lemma 2.6 holds for almost all $\omega \in \mathbf{R}^n$.

To prove Lemma 2.6 in general, we shall show that both polytopes have the same vertices. Let ω' be a generic vector in $\mathbf{R}^n$. Then, for $\epsilon > 0$ sufficiently small,

$$\begin{aligned} face_{\omega'}\big(face_\omega\big(State_d(I)\big)\big) &=_{(2.3)} face_{\omega+\epsilon\cdot\omega'}\big(State_d(I)\big) \\ &= State_d\big(\, in_{\omega+\epsilon\cdot\omega'}(I)\,\big) =_{(1.3)} State_d\big(\, in_{\omega'}(in_\omega(I))\,\big) \\ &= face_{\omega'}\big(State_d(\, in_\omega(I)\,)\big). \end{aligned}$$

Here the second and the fourth equation use the generic case established above. ∎

The exchange lemma argument in the above proof gives the following corollary.

Corollary 2.7. *If $\prec, \prec'$ are two distinct term orders, then*

$$in_{\prec}(I)_d \neq in_{\prec'}(I)_d \quad \textit{implies} \quad \sum in_{\prec}(I)_d \neq \sum in_{\prec'}(I)_d.$$

Proof of Theorem 2.5: We must show that the Gröbner fan $GF(I)$ equals the fan $\mathcal{N}(State(I))$. Note that $GF(I)$ is a polyhedral complex by Proposition 2.4. All faces of a polyhedral complex are determined by the maximal faces. Therefore two polyhedral complexes coincide if and only if their maximal faces coincide. Thus it suffices to show that the maximal (open) cones in $GF(I)$ coincide with the maximal (open) cones in $\mathcal{N}(State(I))$. Equivalently, we must show that $in_\omega(I) = in_{\omega'}(I)$ if and only if $face_\omega(State(I)) = face_{\omega'}(State(I))$, for any two generic vectors ω, ω' in $\mathbf{R}^n$.

Two homogeneous ideals are equal if and only if they agree in each degree. By definition, the integer D appearing in (2.8) is a universal bound for the degrees of minimal generators for these ideals. Therefore

$$in_\omega(I) = in_{\omega'}(I) \quad \Longleftrightarrow \quad in_\omega(I)_d = in_{\omega'}(I)_d \text{ for all } d = 1, 2, \ldots, D. \tag{2.9}$$

Also note that, by (2.4), we have

$$face_\omega(State(I)) \quad = \quad \sum_{d=1}^{D} face_\omega(State_d(I)), \tag{2.10}$$

and similarly for ω'. The "only-if" direction of our claim follows immediately from (2.9), (2.10) and Lemma 2.6. For the converse suppose that $face_\omega(State(I)) = face_{\omega'}(State(I))$. Since ω, ω' are generic, these faces are vertices, and we may replace ω, ω' by term orders. The "if"-direction now follows from Corollary 2.7 and the remark after equation (2.4). ∎

Passing to the indefinite article, we say that a polytope $Q \in \mathbf{R}^n$ is a *state polytope* for a given ideal $I \subset k[\mathbf{x}]$ if it is strongly isomorphic to $State(I)$. In other words, a polytope Q is a state polytope for I if its normal fan $\mathcal{N}(Q)$ equals the Gröbner fan $GF(I)$.

We shall now discuss state polytopes for three families of examples.

Proposition 2.8. *Let f be a homogeneous polynomial and $I = \langle f \rangle$ the principal ideal it generates. Then the Newton polytope $New(f)$ is a state polytope for I.*

Proof: The singleton $\{f\}$ equals the reduced Gröbner basis with respect to any term order. From (2.6') we see that $C[\omega] = \mathcal{N}_{New(f)}(face_\omega(New(f)))$, i.e., the equivalence classes of term orders are the normal cones of the Newton polytope $New(f)$. ∎

In fact, the same argument proves the following stronger statement:

Corollary 2.9. *Let $\mathcal{G}$ be a universal Gröbner basis of I which is a reduced Gröbner basis of I with respect to every term order. Then $\sum_{g \in \mathcal{G}} New(g)$ is a state polytope for I.*

Example 2.10. *(2×2-minors of a 2×3-matrix)* Recall from Example 1.4 that

$$\mathcal{G} \;=\; \{\, x_{11}x_{22} - x_{12}x_{21},\; x_{11}x_{23} - x_{13}x_{21},\; x_{12}x_{23} - x_{13}x_{22} \,\}$$

is the reduced Gröbner basis for every term order. We can thus apply Corollary 2.9. The three Newton polytopes $New(x_{1i}x_{2j} - x_{1j}x_{2i})$ are line segments lying in a two-dimensional linear subspace of $\mathbf{R}^6$. The symmetry group S_3 permutes them. The Minkowski sum of the three segments is a centrally symmetric hexagon. (In particular, $dim(State(I)) = 2$.) ∎

We finally consider an ideal I generated by linear forms (cf. Example 1.5 and Proposition 1.6). Here Proposition 2.9 does not apply, because the set of all circuits is not a reduced Gröbner basis for I. As in Chapter 1 we let d denote the dimension of the space of linear forms in I. We say that a d-subset $\{i_1, \ldots, i_d\}$ of $\{1, \ldots, n\}$ is a *basis* if the $(n-d)$-set $\{x_1, \ldots, x_n\} \backslash \{x_{i_1}, \ldots, x_{i_d}\}$ is linearly independent modulo I. The *matroid polytope* of the linear ideal I is defined as

$$Mat(I) \quad := \quad conv\{e_{i_1} + \cdots + e_{i_d} \ : \ \{i_1, \ldots, i_d\} \text{ basis}\}. \tag{2.11}$$

This polytope plays a fundamental role in combinatorial optimization.

Proposition 2.11. *If I is an ideal generated by linear forms, then its matroid polytope $Mat(I)$ is a state polytope for I.*

Proof: We first note that the degree bound D used in (2.8) is equal to 1. Therefore $State(I) = State_1(I)$. Next we claim that the initial ideals of I are precisely the ideals $\langle x_{i_1}, \ldots, x_{i_d} \rangle$, where $\{i_1, \ldots, i_d\}$ runs over all bases. Clearly, every initial ideal has this form. Conversely, given any ideal of this form, then its own incidence vector $\omega := e_{i_1} + \cdots + e_{i_d}$ selects $in_\omega(I) = \langle x_{i_1}, \ldots, x_{i_d} \rangle$. Therefore the convex hull in (2.11) coincides with the convex hull in (2.7), and we are done. ∎

Exercises:

(1) Show that every complete fan in $\mathbf{R}^2$ is the normal fan of a polytope. Give an example of a complete fan in $\mathbf{R}^3$ which is not the normal fan of a polytope.

(2) Consider the monomial ideals $M_1 = \langle x^3, y^3, x^2y^2 \rangle$ and $M_2 = \langle x^2y, xy^2, x^4, y^4 \rangle$. Does there exist an ideal $I \subset k[x, y]$ and two term orders $\prec_1$ and $\prec_2$ such that $in_{\prec_1}(I) = M_1$ and $in_{\prec_2}(I) = M_2$?

(3) Compute the matroid polytope for the linear ideal in Exercise (3) of Chapter 1.

(4) Let I be an ideal in $k[x_1, \ldots, x_n]$, $Y \subset \{x_1, \ldots, x_n\}$ a subset of the variables, and $I \cap k[Y]$ the corresponding elimination ideal.
   (a) Show that $State(I \cap k[Y])$ appears as a face of $State(I)$.
   (b) Give an example which shows that this face is generally not unique.

(5) Determine the state polytope and Gröbner fan for the ideal of 2×2-minors of a $2 \times m$-matrix of indeterminates.

(6) Explain the relationship between $State(I + \langle x_i \rangle)$ and $State(I)$.

(7) Given two ideals I_1 and I_2 in the polynomial ring $k[\mathbf{x}]$, what is the relationship between the polytopes $State(I_1 \cdot I_2)$ and $State(I_1) + State(I_2)$?

(8) Show that an ideal I is a monomial ideal if and only if $dim(State(I)) = 0$. Give a characterization of all ideals whose state polytope is one-dimensional.

Notes:

The Newton polytope occupies a central place at the crossroads of algebra, geometry and combinatorics. An excellent general reference, especially for polyhedral methods in elimination theory, is (Gel'fand, Kapranov & Zelevinsky 1994). Minkowski addition of polytopes was applied to Gröbner bases in (Gritzmann & Sturmfels 1993). The state polytope of an ideal was introduced in (Bayer & Morrison 1988). The Gröbner fan was introduced in (Mora & Robbiano 1988). A thorough analysis of the polytopes $State_d(I)$ and their normal fans for varying d was undertaken by Mall (1995). A natural Minkowski summand of the state polytope is the *Chow polytope* of an algebraic cycle in projective space. It was shown in (Kapranov, Sturmfels & Zelevinsky 1992) that the Chow polytope can be realized as a suitable limit $lim_{d\to\infty} State_d(I)$.

CHAPTER 3

Variation of Term Orders

In this chapter we discuss algorithmic issues in Gröbner basis theory which involve the variation of term orders for a fixed ideal I. The global "combinatorial space" for studying such variations is the state polytope $State(I)$, or, equivalently, the Gröbner fan $GF(I)$. In what follows we let I denote a fixed ideal in $k[\mathbf{x}]$, and we assume that I is homogeneous with respect to some positive grading $deg(x_i) = d_i > 0$. Our first topic is algorithms for computing the state polytope and a universal Gröbner basis of I.

For the computation of the state polytope $State(I)$, it is useful to have the following easy formula for its dimension. If $\mathcal{F}$ is any finite set of polynomials in $k[\mathbf{x}]$, then we abbreviate the Minkowski sum of their Newton polytopes as follows:

$$New(\mathcal{F}) \quad := \quad \sum_{f \in \mathcal{F}} New(f) \quad = \quad New\,\Big(\prod_{f \in \mathcal{F}} f\Big). \tag{3.1}$$

We call $New(\mathcal{F})$ the *Newton polytope* of the *set* $\mathcal{F}$.

Lemma 3.1. *Let $\mathcal{G}$ be any reduced Gröbner basis for I. Then the affine span of $New(\mathcal{G})$ and the affine span of $State(I)$ are parallel linear subspaces in $\mathbf{R}^n$. In particular,*

$$dim\,\big(\,New(\mathcal{G})\,\big) \quad = \quad dim\,\big(\,State(I)\,\big). \tag{3.2}$$

Proof: Let $\omega \in \mathbf{R}^n$ be a weight vector for the Gröbner basis $\mathcal{G}$. Then $\mathbf{v} := face_\omega(New(\mathcal{G}))$ is a vertex of $New(\mathcal{G})$, and $\mathbf{v}' := face_\omega(State(I))$ is a vertex of $State(I)$. It was shown in equation (2.6') that the normal cone of $\mathbf{v}$ at $New(\mathcal{G})$ is equal to the normal cone of $\mathbf{v}'$ at $State(I)$. This cone is denoted $\overline{C[\omega]}$ (see (2.6)). Consider the *lineality space* of this cone, $\overline{C[\omega]} \cap -\overline{C[\omega]}$. This linear subspace is the orthogonal complement of the affine span of $New(\mathcal{G})$, and it is also the orthogonal complement of the affine span of $State(I)$. ∎

Our first algorithm computes the state polytope $State(I)$ by an incremental method. The basic idea is to construct an increasing sequence of polytopes by computing Gröbner bases for more and more term orders. Each polytope P in this sequence is stored by the list of its vertices and the list of its facets F. Each facet F consists of the following data: an outer normal vector $\omega \in \mathcal{N}_P(F)$ and the list of all vertices of P which lie on F.

Algorithm 3.2. *(Computing the state polytope)*
Input: Generators of a homogeneous ideal I in $k[\mathbf{x}]$. A tentative maximum degree D.
Output: Vertices and facets of the state polytope $State(I)$.
0. Fix an arbitrary tie breaking term order $\prec$.

1. Compute the reduced Gröbner basis $\mathcal{G}$ of I with respect to $\prec$.
2. Let $L \subset \mathbf{R}^n$ be the linear subspace which is parallel to the affine span of the Newton polytope $New(\mathcal{G})$. Fix a basis of L. Set $r := dim(L)$. (All subsequent computations are to be carried out inside L.)
3. Choose $r+1$ vectors $\omega_1, \ldots, \omega_{r+1} \in L$ whose positive span equals L.
4. For $i = 1, 2, \ldots, r+1$ compute the reduced Gröbner basis $\mathcal{G}_i$ for I with respect to $\prec_{\omega_i}$. If the maximum degree D_i of an element in $\mathcal{G}_i$ is larger than D, then replace D by that D_i. For $i = 1, 2, \ldots, r+1$ compute $\mathbf{v}_i := \sum_{d=1}^{D} in_{\prec_{\omega_i}}(I)_d$. This is the sum of the exponent vectors of all non-standard monomials of degree at most D.
5. Let $P := conv\{\mathbf{v}_1, \mathbf{v}_2, \ldots, \mathbf{v}_{r+1}\}$. Let `activefacets` be the set of facets of P, where each facet F is a pair (*set of vertices, normal vector*). Set `passivefacets` $:= \emptyset$.
6. While `activefacets` is non-empty do
   6.1. Pick any facet $F = (V, \omega) \in$ `activefacets`. Remove (V, ω) from `activefacets`.
   6.2. Compute the reduced Gröbner basis $\mathcal{G}$ for I with respect to $\prec_\omega$. If the maximum degree D' of an element in $\mathcal{G}$ is larger than D, then replace D by D' and return to step 3. Otherwise compute $\mathbf{v} := \sum_{d=1}^{D} in_{\prec_\omega}(I)_d$.
   6.3. If $\mathbf{v} \in V$ then set `passivefacets` $:=$ `passivefacets` $\cup \{(V, \omega)\}$.
   6.4. If $\mathbf{v} \notin V$ then compute the new polytope $P := conv(P \cup \{\mathbf{v}\})$ using the *beyond-beneath technique*. Update the sets `activefacets` and `passivefacets` accordingly. (Note: all new facet normals ω are to be chosen in the subspace L.)
7. Output the polytope P. Its set of facets equals `passivefacets`.

This algorithm description is very coarse. A few comments are in place. The choice of $r+1$ positively spanning vectors in step 3 has the effect that the polytope P in step 5 has dimension $r = dim(State(I))$, by Lemma 3.1. This simplifies the subsequent computations because, by restricting to the subspace L, we are dealing with full-dimensional polytopes only. This consideration makes each new normal vector ω computed in step 6.4 unique up to a positive scalar multiple. The *beyond-beneath technique* called upon in step 6.4 is the characterization of the face(t)s of the enlarged polytope $conv(P \cup \{\mathbf{v}\})$ in terms of the face(t)s of P and their relative position (beyond-beneath) to the new point $\mathbf{v}$. The beyond-beneath technique is introduced in (Grünbaum 1967). An algorithmic version can be found in Section 8.4 of (Edelsbrunner 1987). We refer to these two books for details on how to implement step 6.4.

The correctness of Algorithm 3.2 can be seen as follows: At each stage in the procedure, P is a full-dimensional subpolytope of $State(I)$. They fail to be equal if and only if there exists a linear functional ω whose maximum over $State(I)$ is larger than its maximum over P. Here it suffices to consider only ω's which are outer normal vectors to facets of P. Moreover, the facets in `passivefacets` need not be tested, since each of them was already shown to span the supporting hyperplane to a facet of $State(I)$. In summary, the subpolytope P fails to be equal to $State(I)$ if and only if there exists $(V, \omega) \in$ `activefacets` such that $\mathbf{v} \notin V$ in step 6.4. This shows the correctness of Algorithm 3.2.

We remark that if we return from step 6.2 to step 3 then, as it stands, all `activefacets` and all `passivefacets` are discarded. In any practical implementation

Algorithm 3.2 should be reorganized to be less wasteful and thus more efficient.

A byproduct of Algorithm 3.2 is the generation of a universal Gröbner basis $\mathcal{U}$ for the ideal I. We simply define $\mathcal{U}$ as the union of all reduced Gröbner bases $\mathcal{G}_i$ and $\mathcal{G}$ which were computed along the way. In fact, the knowledge of the state polytope $State(I)$ of a homogeneous ideal I is essentially equivalent to the knowledge of a universal Gröbner basis $\mathcal{U}$ for I. We shall present algorithms for converting $State(I)$ into $\mathcal{U}$ and conversely.

Algorithm 3.3. *(From the state polytope to a universal Gröbner basis)*
Input: The state polytope $State(I)$ of a homogeneous ideal I in $k[\mathbf{x}]$.
Output: A universal Gröbner basis $\mathcal{U}$ of I.
1. Let $\mathcal{U} := \emptyset$. Set $P = State(I)$. Fix any tie breaking term order $\prec$.
2. For each vertex $\mathbf{v}$ of P do:
   2.1 Select any vector ω in the open cone $\mathcal{N}_P(\{\mathbf{v}\})$.
   2.2 Compute the reduced Gröbner basis $\mathcal{G}$ of I with respect to $\prec_\omega$.
   2.3 Set $\mathcal{U} := \mathcal{U} \cup \mathcal{G}$.

For the converse direction we shall make use of the following fact.

Proposition 3.4. *Let $\mathcal{U}$ be a universal Gröbner basis of a homogeneous ideal I. Then the state polytope $State(I)$ is a Minkowski summand of the Newton polytope $New(\mathcal{U})$.*

Proof: We must show that the normal fan of $New(\mathcal{U})$ refines the Gröbner fan $GF(I)$. Suppose $\omega, \omega' \in \mathbf{R}^n$ lie in the same cell of the normal fan of $New(\mathcal{U}) = \sum_{p \in \mathcal{U}} New(p)$. Then $in_\omega(p) = in_{\omega'}(p)$ for all $p \in \mathcal{U}$. By Corollary 1.9 and the universal Gröbner basis property of $\mathcal{U}$, this implies $in_\omega(I) = in_{\omega'}(I)$, i.e., ω and ω' lie in the same cell of $GF(I)$. ■

Algorithm 3.5. *(From a universal Gröbner basis to the state polytope)*
Input: A universal Gröbner basis $\mathcal{U}$ of a homogeneous ideal I in $k[\mathbf{x}]$.
Output: The state polytope $State(I)$.
0. Let D be the maximum degree of any element in $\mathcal{U}$.
1. Compute the Minkowski sum of the Newton polytopes:

$$New(\mathcal{U}) = \sum_{p \in \mathcal{U}} New(p).$$

2. For each vertex $\mathbf{v}$ of $New(\mathcal{U})$ do:
   2.1 Select any vector ω in the open cone $\mathcal{N}_{New(\mathcal{U})}(\{\mathbf{v}\})$.
   2.2 Read off the initial monomial ideal $in_\omega(I) = \langle in_\omega(p) \, : \, p \in \mathcal{U} \rangle$.
   2.3 Compute the corresponding vertex of the state polytope: $\sum_{d=1}^{D} in_\omega(I)_d$. If this vertex has not been computed previously, then output it now.

The correctness of Algorithm 3.5 follows from Proposition 3.4. Again, what is written here is only a crude outline. For instance, it is a non-trivial issue how to implement step 1. Techniques for computing Minkowski sums can be found in (Gritzmann & Sturmfels 1993). If step 1 computes all faces of P, not just the vertices, then the same can be done for the state polytope in step 2. The point is that the normal fan of $State(I)$ is obtained from the normal fan of P by identifying certain adjacent cones. An alternative way of performing the task of Algorithm 3.5

is to run Algorithm 3.2 on the input $\mathcal{U}$. In that case the Gröbner basis computations in steps 1, 4 and 6.2 are redundant, since $\mathcal{U}$ is already a universal Gröbner basis. It would be interesting to compare these two approaches experimentally.

Algorithms 3.2 and 3.5 for computing the state polytope are global in the sense that they make little use of the local polyhedral information contained in each reduced Gröbner basis. An alternative algorithm is to construct the state polytope by a local search along its edge graph. Here is the general scheme for carrying out such a search.

Algorithm 3.6. *(Computing the state polytope by searching its edge graph)*
Input: Generators of a homogeneous ideal I in $k[\mathbf{x}]$.
Output: Vertices and edges of the state polytope $State(I)$.

1. Choose any random vector $\omega \in \mathbf{R}^n$ and compute the reduced Gröbner basis $\mathcal{G}$ of I with respect to ω. Represent the monomial ideal $in_\omega(I)$ by its minimal generators.
2. Set `Vertices` $:= \{in_\omega(I)\}$ and set `Edges` $:= \emptyset$. Fix an infinitesimal real $\epsilon > 0$.
3. Let $C[\omega]$ be the normal cone at the vertex of the Newton polytope $New(\mathcal{G})$ supported by ω (cf. (2.6)). Let $F_1, F_2, \ldots, F_s$ denote the facets of $C[\omega]$. Select a vector ω_i in the relative interior of each facet F_i.
4. For i from 1 to s do
   4.1. Let $\omega' := \omega_i - \epsilon \cdot \omega$.
   4.2. Transform $\mathcal{G}$ into the reduced Gröbner basis $\mathcal{G}'$ of I with respect to ω'.
   4.3. `Edges` $:=$ `Edges` $\cup \big\{\{in_\omega(I), in_{\omega'}(I)\}\big\}$.
   4.4. If $in_{\omega'}(I) \notin$ `Vertices` then
      4.4.1. `Vertices` $:=$ `Vertices` $\cup \big\{in_{\omega'}(I)\big\}$
      4.4.2. Proceed recursively by calling step 3 of this algorithm, with ω replaced by ω' and $\mathcal{G}$ replaced by $\mathcal{G}'$.
5. Let D be the maximum degree of any minimal generator of an ideal in `Vertices`.
6. For every monomial ideal M in `Vertices` do
   6.1. Output M and the corresponding vertex of the state polytope, $\sum_{d=1}^{D} M_d$.
   6.2. If desired, output all edges in `Edges` which contain M.

Algorithm 3.6 searches the edge graph of the state polytope indirectly, namely, by traversing the maximal cells in the Gröbner fan. The key ingredient of this algorithm is step 4.2, the transformation from a reduced Gröbner basis to a neighboring reduced Gröbner basis. This step will be described in Subroutine 3.7 below. It uses the following notation and assumptions. Let C_1 and C_2 be two open cells in the Gröbner fan, let $\omega_1 \in C_1$ and $\omega_2 \in C_2$, and let $\omega \in \overline{C_1} \cap \overline{C_2}$.

Subroutine 3.7. *(Local change of reduced Gröbner bases)*
Input: The reduced Gröbner basis $\mathcal{G}_1$ of I with respect to ω_1.
Output: The reduced Gröbner basis $\mathcal{G}_2$ of I with respect to ω_2.

1. Let $\mathcal{H}_1 := in_\omega(\mathcal{G}_1) = \{in_\omega(g) : g \in \mathcal{G}_1\}$.
   (This is the reduced Gröbner basis of the ideal $in_\omega(I)$ with respect to ω_1.)
2. Compute the reduced Gröbner basis $\mathcal{H}_2$ of $in_\omega(I) = \langle \mathcal{H}_1 \rangle$ with respect to ω_2.
3. Set $\mathcal{G}_2' := \emptyset$.
4. For each h in $\mathcal{H}_2$ do

4.1. Reduce h to zero modulo $\mathcal{H}_1$ using the term order ω_1, and keep track of the coefficient polynomials during the reduction. This gives an ω-homogeneous representation

$$h = \sum_{g \in \mathcal{G}_1} p_g \cdot in_\omega(g). \tag{3.3}$$

4.2. Compute the polynomial

$$\sum_{g \in \mathcal{G}_1} p_g \cdot g, \tag{3.4}$$

and add it to the set $\mathcal{G}_2'$.

(After this loop the set $\mathcal{G}_2'$ is a minimal Gröbner basis for I with respect to ω_2.)

5. Transform the minimal Gröbner basis $\mathcal{G}_2'$ into the reduced Gröbner basis $\mathcal{G}_2$.

Proof of correctness of Subroutine 3.7: Since ω_2 represents a term order which refines ω for I, we may assume that ω_2 is arbitrarily close to ω (by replacing ω_2 by $\omega + \epsilon\omega_2$). We must show that $\mathcal{G}_2'$ is a minimal Gröbner basis for I with respect to ω_2. This means that $in_{\omega_2}(\mathcal{G}_2') = \{in_{\omega_2}(f) : f \in \mathcal{G}_2'\}$ minimally generates the monomial ideal $in_{\omega_2}(I) = in_{\omega_2}(in_\omega(I))$. Since $\mathcal{H}_2$ is the reduced Gröbner basis for $in_\omega(I)$ with respect to ω_2, it suffices to show that $in_{\omega_2}(\mathcal{H}_2) = in_{\omega_2}(\mathcal{G}_2')$. Let $f \in \mathcal{G}_2'$ and consider its representation (3.4). The expression (3.3) being ω-homogeneous means that each monomial in the expansion of h has the same ω-weight. Therefore h is the initial form of f with respect to ω. By the above closeness assumption, we have $in_{\omega_2}(f) = in_{\omega_2}(in_\omega(f)) = in_{\omega_2}(h)$, as desired. ∎

Example 3.8. *(Two ternary quadrics)*
Consider the generic complete intersection

$$\begin{aligned} I \;&=\; \langle\, a_1x^2 + a_2xy + a_3xz + a_4y^2 + a_5yz + a_6z^2, \\ &\quad\;\; b_1x^2 + b_2xy + b_3xz + b_4y^2 + b_5yz + b_6z^2 \,\rangle, \end{aligned}$$

where $k = \mathbf{Q}(a_1, \ldots, a_6, b_1, \ldots, b_6)$ is the field of rational functions in the 12 indeterminate coefficients. Choose the weight vector $\omega_1 = (3, 2, 1)$. Then $in_{\omega_1}(I) = \langle x^2, xy, y^3\rangle$. The corresponding reduced Gröbner basis $\mathcal{G}_1 = \{g_1, g_2, g_3\}$ of I has the structure

$$\begin{aligned} g_1 &= \underline{\alpha_1x^2} + \alpha_2y^2 + \alpha_3xz + \alpha_4yz + \alpha_5z^2, \\ g_2 &= \underline{\beta_1xy} + \beta_2y^2 + \beta_3xz + \beta_4yz + \beta_5z^2, \\ g_3 &= \underline{\gamma_1y^3} + \gamma_2xz^2 + \gamma_3y^2z + \gamma_4yz^2 + \gamma_5z^3. \end{aligned}$$

The corresponding cone in the Gröbner fan equals

$$C[\omega_1] \;=\; \{\, (v_1, v_2, v_3) \in \mathbf{R}^3 \;:\; v_1 > v_2 \text{ and } 3v_2 > v_1 + 2v_3 \}.$$

The lineality space of $C[\omega_1]$ is one-dimensional, hence

$$dim(State(I)) = dim(\mathcal{N}(\mathcal{G}_1)) = 2.$$

We now demonstrate the change of Gröbner bases across the facet $\{3v_2 = v_1 + 2v_3\}$. Select a vector in the relative interior of this facet, say $\omega = (3, 1, 0)$.

Then $\mathcal{H}_1 = in_\omega(\mathcal{G}_1) = \{\alpha_1 x^2, \beta_1 xy, \gamma_1 y^3 + \gamma_2 xz^2\}$ is the reduced Gröbner basis of $in_\omega(I)$ with respect to ω_1. (We don't bother cancelling initial coefficients.) Now set $\omega_2 := \omega - \epsilon\omega_1$ for $\epsilon > 0$ very small. The reduced Gröbner basis of $in_\omega(I)$ with respect to ω_2 is found to be $\mathcal{H}_2 = \mathcal{H}_1 \cup \{h\}$,

$$\text{where} \qquad h \quad = \quad \beta_1\gamma_1 y^4 \quad = \quad \beta_1 y \cdot in_\omega(g_3) \; - \; \gamma_2 z^2 \cdot in_\omega(g_2).$$

This is the representation (3.3) in Subroutine 3.7. The new polynomial (3.4) equals

$$\begin{aligned} f \quad &= \quad \beta_1 y \cdot g_3 \; - \; \gamma_2 z^2 \cdot g_2 \\ &= \quad \underline{\beta_1\gamma_1 y^4} - \beta_3\gamma_2 xz^3 + \beta_1\gamma_3 y^3 z + (\beta_1\gamma_4 - \beta_2\gamma_2) y^2 z^2 \\ &\quad + (\beta_1\gamma_5 - \beta_4\gamma_2) yz^3 - \beta_5\gamma_2 z^4. \end{aligned}$$

Therefore $\mathcal{G}_1 \cup \{f\}$ is a minimal Gröbner basis for I with respect to ω_2. However, it is not reduced yet, since the term $-\beta_3\gamma_2 xz^3$ of f is divisible by $in_{\omega_2}(g_3) = \gamma_2 xz^2$. We perform the corresponding reduction to get

$$\begin{aligned} f' = f + \beta_3 z \cdot g_3 = \beta_1\gamma_1 y^4 \; &+ \; (\beta_1\gamma_3 + \beta_3\gamma_1) y^3 z \; + \; (\beta_1\gamma_4 - \beta_2\gamma_2 + \beta_3\gamma_3) y^2 z^2 \\ &+ \; (\beta_1\gamma_5 - \beta_4\gamma_2 + \beta_3\gamma_4) yz^3 \; + \; (\beta_3\gamma_5 - \beta_5\gamma_2) z^4. \end{aligned}$$

We conclude that the reduced Gröbner basis of I with respect to ω_2 equals $\mathcal{G}_2 = \mathcal{G}_1 \cup \{f'\}$. We read off the corresponding cone in the Gröbner fan:

$$C[\omega_2] \quad = \quad \big\{ (v_1, v_2, v_3) \in \mathbf{R}^3 \; : \; v_2 > v_3 \text{ and } v_1 + 2v_3 > 3v_2 \big\}.$$

The two initial ideals we have found so far cover all term orders with $x > y > z$ since

$$\overline{C[\omega_1]} \cup \overline{C[\omega_2]} \quad = \quad \{ (v_1, v_2, v_3) \in \mathbf{R}^3 \; : \; v_1 \geq v_2 \geq v_3 \}.$$

Since the given complete intersection was generic, we can argue by symmetry that there are two distinct initial ideals for each of the six permutations of the variables x, y, z. In summary: *The state polytope of the ideal of two generic ternary quadrics is a planar 12-gon.* ∎

By repeated application of Subroutine 3.7, we can transform the reduced Gröbner basis $\mathcal{G}$ of an ideal I with respect to some term order $\prec$ into the reduced Gröbner basis $\mathcal{G}'$ with respect to any other term order $\prec'$. This is accomplished as follows: Select weight vectors $\omega, \omega' \in \mathbf{R}^n$ which represent $\prec$ and $\prec'$ respectively. Now move the weight vector along the line segment $\{\lambda\cdot\omega' + (1-\lambda)\cdot\omega \; : \; 0 \leq \lambda \leq 1\}$ by incrementally increasing λ from 0 to 1. Each time we pass through a codimension 1 face of the Gröbner fan $GF(I)$, Subroutine 3.7 is called to update the given Gröbner basis. This method for transforming Gröbner bases was proposed by Collart, Kalkbrener & Mall (1996). It is a polyhedral alternative to the more direct algorithm of Faugére, Gianni, Lazard & Mora (1992). An important application of their technique is to change reverse lexicographic Gröbner bases into lexicographic ones, as the latter are much harder to compute than the former.

We next address the question of how to select the "best" term order for a given system of polynomials. As a motivation for this question, consider the following simple example.

Example 3.9. Suppose we wish to find all complex zeros of the polynomial system

$$\mathcal{F} \;=\; \{\, x_1^5 + x_2^3 + x_3^2 - 1 \,,\;\; x_1^2 + x_2^2 + x_3 - 1 \,,\;\; x_1^6 + x_2^5 + x_3^3 - 1 \,\} \quad \subset \quad \mathbf{Q}[x_1, x_2, x_3].$$

A first attempt might be to compute a purely lexicographic Gröbner basis. There are six lexicographic Gröbner bases, one for each ordering of the variables. Unfortunately, each of the six lexicographic Gröbner bases for $\langle \mathcal{F} \rangle$ contains high degree polynomials with very large coefficients. For instance, a typical polynomial in the lexicographic Gröbner basis induced by $x_1 \prec x_2 \prec x_3$ has degree 21 and the maximal appearing integer coefficient is 1553067597584776499. This means that the Gröbner basis computation is rather slow.

For this example there is a much better term order, namely, $\omega = (3,4,7)$. For these weights, our input set $\mathcal{F}$ has the initial terms x_1^5, x_2^2 and x_3^3. These monomials are relatively prime, and hence $\mathcal{F}$ is already a Gröbner basis. We see that the ideal $\langle \mathcal{F} \rangle$ is zero-dimensional and has $30 = 5 \cdot 2 \cdot 3$ zeros up to multiplicities in affine 3-space $\mathbf{C}^3$. ∎

The lucky term order $\omega = (3,4,7)$ in Example 3.9 is found systematically as follows.

Algorithm 3.10. *(Gröbner basis detection)*
Input: A set $\mathcal{F} \subset k[\mathbf{x}]$ of polynomials.
Output: A term order $\omega \in \mathbf{R}^n$ such that $\mathcal{F}$ is a Gröbner basis with respect to ω, if such ω exists; "NO" otherwise.

1. Compute the Newton polytope $P = New(\mathcal{F})$.
2. For each vertex $\mathbf{v}$ of P do:
   - 2.1. Decide whether the cone $\mathcal{N}_P(\{\mathbf{v}\})$ intersects the positive orthant $\mathbf{R}^n_+$. If yes, then
     - 2.1.1. Select $\omega \in \mathcal{N}_P(\{\mathbf{v}\}) \cap \mathbf{R}^n_+$.
     - 2.1.2. Using the Buchberger criterion, decide whether $\mathcal{F}$ is a Gröbner basis with respect to the choice of initial terms defined by ω. If yes, then output ω.
3. If the answer in 2.1.2 has never been "yes", then output "NO".

For most polynomial systems arising in practise, the output of Algorithm 3.10 will undoubtedly be "NO". Nevertheless the same procedure can be used to determine the "best" term order with respect to any predefined local criterion. In what follows we define a natural such criterion. Fix a set of polynomials $\mathcal{F} \subset k[\mathbf{x}]$. For each weight vector arising in step 2.1.1 above, we consider the monomial ideal

$$M_\omega \quad := \quad \langle\, in_\omega(\mathcal{F}) \,\rangle \quad = \quad \langle\, in_\omega(f) \,:\, f \in \mathcal{F} \,\rangle.$$

We define $h_\omega(r)$ to be the Hilbert polynomial of the homogeneous ideal M_ω, that is,

$$h_\omega(r) \quad = \quad dim_k\big(\, (k[\mathbf{x}]/M_\omega)_r \,\big) \qquad \text{for } r \gg 0.$$

If ω and ω' are two generic weight vectors as in 2.1.1, then we say that ω is *better* than ω' if the initial coefficient of the polynomial $h_\omega(r) - h_{\omega'}(r)$ is negative. With this criterion (and a fast subroutine for computing Hilbert polynomials), Algorithm 3.10 can be used to generate a short list of *best* term orders. Our criterion is justified by the following result.

Proposition 3.11. *Let $\mathcal{F}$ be a homogeneous set of polynomials and $\omega \in \mathbf{R}^n$ such that $\mathcal{F}$ is a Gröbner basis with respect to ω. Then ω is a best term order for $\mathcal{F}$.*

Proof: The ideal M_ω is a subideal of $in_\omega(\langle\mathcal{F}\rangle)$. Equality holds if and only if $\mathcal{F}$ is a Gröbner basis with respect to ω. In particular, $h_\omega(r)$ is an upper bound for the number of ω-standard monomials of large degree r modulo $\langle\mathcal{F}\rangle$, and equality holds in the Gröbner basis case. ∎

Our last topic in this chapter is a characterization of term orders by their reduction properties. By a *marked polynomial* we mean a polynomial $f \in k[\mathbf{x}]$ together with a specified *initial term* $in(f)$. Here $in(f)$ can be any of the terms appearing in f. Given a set $\mathcal{F}$ of marked polynomials, we define the *reduction relation modulo* $\mathcal{F}$ in the usual sense of Gröbner bases. We say that $\mathcal{F}$ is marked *coherently* if there exists a term order $\prec$ on $k[\mathbf{x}]$ such that $in(f) = in_\prec(f)$ for all f in $\mathcal{F}$. Clearly, if $\mathcal{F}$ is marked coherently, then the reduction relation "$\rightarrow_\mathcal{F}$" is Noetherian. The following theorem establishes the converse.

Theorem 3.12. *A finite set $\mathcal{F} \subset k[\mathbf{x}]$ of marked polynomials is marked coherently if and only if the reduction relation modulo $\mathcal{F}$ is Noetherian, i.e., every sequence of reductions modulo $\mathcal{F}$ terminates.*

The following example of an incoherent marking was given in Example 1.4, case (3):

$$\underline{x_{11}x_{22}} - x_{12}x_{21}, \qquad \underline{x_{13}x_{21}} - x_{11}x_{23}, \qquad \underline{x_{12}x_{23}} - x_{13}x_{22}. \tag{3.6}$$

Theorem 3.12 tells us that the corresponding reduction relation is not Noetherian. For instance, the following reduction sequence modulo the marked polynomials (3.6) is infinite

$$\begin{aligned} x_{11}x_{12}x_{13}x_{21}x_{22}x_{23} \;\rightarrow\; x_{12}^2x_{13}x_{21}^2x_{23} \;&\rightarrow\; x_{11}x_{12}^2x_{21}x_{23}^2 \\ &\rightarrow\; x_{11}x_{12}x_{13}x_{21}x_{22}x_{23} \;\rightarrow\; \cdots \end{aligned}$$

To prove Theorem 3.12 in general, we need to establish the following lemma.

Lemma 3.13. *Let $\mathcal{F} = \{f_1, \ldots, f_l\} \subset k[\mathbf{x}]$ be marked incoherently. Then there exists a reduction sequence modulo $\mathcal{F}$ which does not terminate.*

Proof: Let $f_i = \mathbf{x}^{\alpha_i} - c_1\mathbf{x}^{\alpha_i+\gamma_{i1}} - c_2\mathbf{x}^{\alpha_i+\gamma_{i2}} - \cdots - c_{s_i}\mathbf{x}^{\alpha_i+\gamma_{is_i}}$, where $\gamma_{ij} = (\gamma_{ij1}, \ldots, \gamma_{ijn})$ are distinct, non-zero vectors in $\mathbf{Z}^n$. Suppose that the marking $in(f_i) = \mathbf{x}^{\alpha_i}$ is incoherent, i.e., there does not exist a term order $\prec$ on $k[\mathbf{x}]$ such that $in_\prec(f_i) = \mathbf{x}^{\alpha_i}$ for all $i = 1, \ldots, l$.

Let $e_1, \ldots, e_n$ denote the standard coordinate vectors in $\mathbf{Q}^n$. Since $\mathcal{F}$ is marked incoherently, the following system of linear inequalities is infeasible (cf. Proposition 1.11):

$$\omega^T \cdot \left[e_1, \ldots, e_n, -\gamma_{11}, \ldots, -\gamma_{1s_1}, \ldots, -\gamma_{ij}, \ldots, -\gamma_{l1}, \ldots, -\gamma_{ls_l}\right] \quad > \quad 0.$$

Here the γ_{ij} are column vectors. By Linear Programming Duality (Schrijver 1986, Section 7.3), there exists a non-zero, non-negative integer vector

$$y = (\tilde{y}_1, \ldots, \tilde{y}_n, y_{11}, \ldots, y_{1s_1}, \ldots, y_{ij}, \ldots, y_{l1}, \ldots, y_{ls_l})$$

such that

$$[e_1, \ldots, e_n, -\gamma_{11}, \ldots, -\gamma_{1s_1}, \ldots, -\gamma_{ij}, \ldots, -\gamma_{l1}, \ldots, -\gamma_{ls_l}] \cdot y \quad = \quad 0. \qquad (3.6)$$

In particular, $y_{11}\gamma_{11} + \ldots + y_{ls_l}\gamma_{ls_l}$ is a non-negative vector. We say that a solution vector y to (3.6) is *minimal* if $N := \sum_{i=1}^{l} \sum_{j=1}^{s_i} y_{ij}$ is as small as possible. Among the possible solutions to (3.6), if there exists a solution y with $y_{11}\gamma_{11} + \ldots + y_{ls_l}\gamma_{ls_l} = 0$, we choose a minimal such solution. If all solutions y are such that $y_{11}\gamma_{11} + \ldots + y_{ls_l}\gamma_{ls_l} > 0$ we choose a minimal solution among *all* solutions to (3.6).

We will construct an infinite reduction sequence modulo $\mathcal{F}$. Let $\mathbf{x}^{\beta} = x_1^{\beta_1} \cdots x_n^{\beta_n} := \prod_{i=1}^{l} in(f_i)^{\nu_i}$, where $\nu_i = y_{i1} + \cdots + y_{is_i}$, and let $D = \sum_{i=1}^{l} s_i$. We choose a sequence of non-negative vectors $p_0, p_1, \ldots, p_N$ in $\mathbf{Z}^D$, where $p_r = (p_{r1}, \ldots, p_{rl})$ and $p_{ri} = (p_{ri1}, \ldots, p_{ris_i}) \in \mathbf{Z}^{s_i}$ according to the following rules:

(a) For $r = 0, 1, 2, \ldots, N$ we require $p_{r11} + \cdots + p_{r1s_1} + \ldots + p_{rl1} + \ldots + p_{rls_l} = r$.

(b) We require the componentwise inequalities $0 = p_0 \leq p_1 \leq \ldots \leq p_{N-1} \leq p_N = y$.

For each $r = 0, 1, 2, \ldots, N$ we define

$$\beta^{(r)} \quad := \quad \beta + p_{r1}\gamma_1 + p_{r2}\gamma_2 + \ldots + p_{rl}\gamma_l,$$

where $\gamma_i = (\gamma_{i1}, \ldots, \gamma_{is_i})$. Our construction implies that all $\beta^{(r)}$ are non-negative integer vectors and that the monomial $\mathbf{x}^{\beta}$ divides the monomial $\mathbf{x}^{\beta^{(N)}}$. To see this, note $\beta = \sum_i \sum_j y_{ij}\alpha_i$, and $\beta^{(N)} = \sum_i \sum_j y_{ij}(\alpha_i + \gamma_{ij})$, and at each intermediate step $\beta^{(r)}$ one of the γ_{ij} enters the double sum.

For each $r = 1, 2, \ldots, N$, the vector $p_r - p_{r-1} \in \mathbf{Z}^D$ consists of a unique entry 1 and 0's elsewhere. The entry 1 is given by two indices $i_r \in \{1, \ldots, l\}$ and $j_r \in \{1, \ldots, s_{i_r}\}$. We define the set of *stage r monomials* as $stage(r) := \{\mathbf{x}^{\beta^{(r-1)} + \gamma_{i_r 1}}, \ldots, \mathbf{x}^{\beta^{(r-1)} + \gamma_{i_r s_{i_r}}}\}$.

Let $red(\mathbf{x}^{\beta^{(r)}}, f_{i_r})$ denote the polynomial obtained by reducing $\mathbf{x}^{\beta^{(r-1)}}$ once with respect to f_{i_r}. The monomials appearing in $red(\mathbf{x}^{\beta^{(r)}}, f_{i_r})$ are precisely the monomials in $stage(r)$. (Also note that $\mathbf{x}^{\beta^{(r)}}$ lies in $stage(r)$ since the conditions (a) and (b) imply $\beta^{(r)} = \beta^{(r-1)} + \gamma_{i_r j_r}$ for some $1 \leq j_r \leq s_{i_r}$). Consider the sequence of reductions $g_0 \to_{\mathcal{F}} g_1 \to_{\mathcal{F}} g_2 \to_{\mathcal{F}} \ldots \to_{\mathcal{F}} g_N$ which is obtained by reducing $\mathbf{x}^{\beta^{(r-1)}}$ in the r-th step by f_{i_r}, where the index i_r is determined as above by the sequence of p_r's. More precisely, we define

$$\begin{aligned}
g_0 &:= \mathbf{x}^{\beta} \\
g_1 &:= red(\mathbf{x}^{\beta}, f_{i_1}), \\
g_2 &:= g_1 - coef(\mathbf{x}^{\beta^{(1)}}, g_1)(\mathbf{x}^{\beta^{(1)}} - red(\mathbf{x}^{\beta^{(1)}}, f_{i_2})), \\
&\vdots \\
g_r &:= g_{r-1} - coef(\mathbf{x}^{\beta^{(r-1)}}, g_{r-1})(\mathbf{x}^{\beta^{(r-1)}} - red(\mathbf{x}^{\beta^{(r-1)}}, f_{i_r})).
\end{aligned}$$

Here $coef(\mathbf{x}^A, h)$ is the coefficient of $\mathbf{x}^A$ in h. Note that g_r is a polynomial involving only monomials in the set $\bigcup_{j=1}^{r} stage(j)$. In order to prove Lemma 3.13, we need to show that no unwanted cancellations occur. The following lemma is the key.

Lemma 3.14. *For $1 \le r, t \le N$, the monomial $\mathbf{x}^{\beta^{(r)}}$ lies in stage(t) if and only if $r = t$.*

Proof: Suppose that $\mathbf{x}^{\beta^{(r)}} \in stage(t)$ for $t \neq r$. This implies

$$\begin{aligned} \beta^{(r)} &= \beta + p_{r1}\gamma_1 + \ldots + p_{rl}\gamma_l \\ &= \beta + p_{t-1,1}\gamma_1 + \ldots + p_{t-1,l}\gamma_l + \gamma_{i_t j} = \beta^{(t-1)} + \gamma_{i_t j}, \end{aligned}$$

for some $j \in \{1, \ldots, s_{i_t}\}$. We define a vector in $\mathbf{Z}^d$ as follows:

$$y' := \begin{cases} p_{t-1} + e_{i_t j} - p_r & \text{if } t > r \\ y - p_r + p_{t-1} + e_{i_t j} & \text{if } t < r \,. \end{cases}$$

By the requirements (a) and (b), the vector $y' = (y'_{11}, \ldots, y'_{1s_1}, \ldots, y'_{l1}, \ldots, y'_{ls_l})$ is non-zero and non-negative. Also, $\sum_{i=1}^{l}\sum_{j=1}^{s_i} y'_{ij} = \begin{cases} t - r & \text{if } t > r \\ N - r + t & \text{if } t < r \end{cases}$, and consequently $\sum_{i=1}^{l}\sum_{j=1}^{s_i} y'_{ij} < N$. The vector $y' \cdot \gamma = y'_1\gamma_1 + \ldots + y'_l\gamma_l \in \mathbf{Z}^n$ is either zero (if $t > r$) or equal to $y \cdot \gamma := y_1\gamma_1 + \cdots + y_l\gamma_l$ (if $t < r$). In particular, if $y \cdot \gamma = 0$, then $y' \cdot \gamma = 0$ as well. In any case, $y' \cdot \gamma$ is non-negative, and therefore the vector $y' \in \mathbf{N}^D$ can be completed to a solution $(\tilde{y}', y')$ of (3.6). But y was chosen to be minimal among such solutions to (3.6), hence this is a contradiction. ∎

Proof of Lemma 3.13 continued: Lemma 3.14 implies that $coef(\mathbf{x}^{\beta^{(r)}}, g_r) \neq 0$ for $r = 1, 2, \ldots, N$. At the last stage in our reduction sequence we get $g_N = c\,\mathbf{x}^{\beta^{(N)}} + g$ where c is non-zero and, again by Lemma 3.14, all monomials of g are contained in $\bigcup_{j=1}^{N} stage(j) \setminus \{\mathbf{x}^{\beta^{(0)}}, \ldots, \mathbf{x}^{\beta^{(N)}}\}$. Since $\beta^{(N)} - \beta = \sum y_i\gamma_i$ is non-negative, we can start the reduction sequence again, with $\beta^{(N)}$ replacing β. If $\beta^{(N)} = \beta$ (that is, $y \cdot \gamma = 0$), then this reduction sequence takes the form

$$\begin{aligned} \mathbf{x}^{\beta} \;&\longrightarrow_{\mathcal{F}}\; c\,\mathbf{x}^{\beta} + g \;\longrightarrow_{\mathcal{F}}\; c^2\mathbf{x}^{\beta} + (1 + c)\,g \\ &\longrightarrow_{\mathcal{F}}\; c^3\mathbf{x}^{\beta}(1 + c + c^2)\,g \;\longrightarrow_{\mathcal{F}}\; \cdots \end{aligned}$$

and is clearly infinite.

If, on the other hand, the non-negative vector $y \cdot \gamma = \beta^{(N)} - \beta$ is non-zero, then this reduction sequence takes the form

$$\begin{aligned} \mathbf{x}^{\beta} \;&\longrightarrow_{\mathcal{F}}\; c\,\mathbf{x}^{\beta^{(N)}} + g \;\longrightarrow_{\mathcal{F}}\; c^2\mathbf{x}^{2\beta^{(N)}-\beta} + \left(1 + c\,\mathbf{x}^{\beta^{(N)}-\beta}\right) g \\ &\longrightarrow_{\mathcal{F}}\; c^3\mathbf{x}^{3\beta^{(N)}-2\beta} + \left(1 + c\mathbf{x}^{\beta^{(N)}-\beta} + c^2\mathbf{x}^{2\beta^{(N)}-2\beta}\right) g \;\longrightarrow_{\mathcal{F}}\; \cdots \end{aligned}$$

To see that this sequence is infinite as well, we must show that no unwanted cancellations can occur. An unwanted cancellation would imply that an identity $\beta^{(r)} = \beta^{(t-1)} + \gamma_{i_t j}$ holds modulo the integer linear span of $\beta^{(N)} - \beta = y \cdot \gamma$. To show that this is impossible, we shall call upon our minimality assumption for the solution y once more.

Suppose there exists an integer $m > 0$ such that $\beta^{(r)} + m(y \cdot \gamma) = \beta^{(t-1)} + \gamma_{i_t j}$. (The case $\beta^{(r)} = m(y \cdot \gamma) + \beta^{(t-1)} + \gamma_{i_t j}$ is analogous.) We define

$$y' := \begin{cases} p_{t-1} + e_{i_t j} - p_r & \text{if } t > r \\ my - p_r + p_{t-1} + e_{i_t j} & \text{if } t < r \,. \end{cases}$$

Note that as before, $y' \in \mathbf{Z}^D$ is a non-negative, non-zero solution to (3.6). For $t > r$, as in Lemma 3.14, the vector y' is a smaller solution to (3.6), contradicting our choice of y. For $t < r$, we have $y' \cdot \gamma = 0$, again contradicting our choice of y. This completes the proof of Lemma 3.13 and of Theorem 3.12. ∎

Exercises:

(1) Generalizing Example 3.9, consider the polynomial system

$$\begin{aligned}\mathcal{F} &= \{x_1^{a_1} + x_2^{a_2} + x_3^{a_3} - 1,\ x_1^{b_1} + x_2^{b_2} + x_3^{b_3} - 1,\ x_1^{c_1} + x_2^{c_2} + x_3^{c_2} - 1\} \\ &\subset \mathbf{Q}[x_1, x_2, x_3].\end{aligned}$$

Give necessary and sufficient conditions on the exponents $a_1, a_2, a_3, b_1, b_2, b_3, c_1, c_2, c_3$ such that $\mathcal{F}$ is a Gröbner basis with respect to some term order.

(2) Let I be the ideal generated by two generic linear forms in $k[x_1, x_2, x_3, x_4]$, and let $\mathcal{U}$ be the universal Gröbner basis consisting of the six circuits of I (cf. Proposition 1.6). Compute $New(\mathcal{U})$ and $State(I)$ explicitly, and verify Proposition 3.4 in this case.

(3) In Corollary 2.9 a sufficient condition was given for the equality $New(\mathcal{U}) = State(I)$ to hold in Proposition 3.4. Is this condition also necessary ?

(4) Compute the state polytope of the ideal of two generic ternary cubics.

(5) List all coherent markings for the set of 3×3-minors of a 3×5-matrix of indeterminates.

(6) Consider the Grassmann variety $Grass_{r,s}$ of r-dimensional linear subspaces in k^s. What is the dimension of the state polytope of its vanishing ideal in the Plücker embedding ?

Notes:

Subsequent to the fundamental work on this subject in (Bayer 1982) and (Bayer & Morrison 1988), Bayer and Morrison implemented an algorithm for computing the state polytope of an ideal. The emphasis of their work was on space curves with a view toward applications in geometric invariant theory. A new implementation was done in 1991-92 by Alyson Reeves (unpublished) and applied to the study of Borel-fixed ideals. Reeves' program for computing the state polytope is available in (some versions of) MACAULAY under the command `hull`. It is based on Algorithm 3.2. Algorithm 3.6 has not yet been fully implemented. Subroutine 3.7 for locally changing Gröbner bases is due to Collart, Kalkbrener & Mall (1996). Earlier work on transforming Gröbner bases from one term order to another term order is found in (Faugére, Gianni, Lazard & Mora 1992). Algorithm 3.10 and Proposition 3.11 appeared in (Gritzmann & Sturmfels 1993). Theorem 3.12 and its proof are taken from (Reeves & Sturmfels 1993).

CHAPTER 4

Toric Ideals

We shall study a special class of ideals in $k[\mathbf{x}] := k[x_1, \ldots, x_n]$. Fix a subset $\mathcal{A} = \{\mathbf{a}_1, \ldots, \mathbf{a}_n\}$ of $\mathbf{Z}^d$. Each vector $\mathbf{a}_i$ is identified with a monomial $\mathbf{t}^{\mathbf{a}_i}$ in the Laurent polynomial ring $k[\mathbf{t}^{\pm 1}] := k[t_1, \ldots, t_d, t_1^{-1}, \ldots, t_d^{-1}]$. Consider the semigroup homomorphism

$$\pi : \mathbf{N}^n \rightarrow \mathbf{Z}^d, \quad \mathbf{u} = (u_1, \ldots, u_n) \mapsto u_1\mathbf{a}_1 + \cdots + u_n\mathbf{a}_n. \tag{4.1}$$

The image of π is the semigroup

$$\mathbf{N}\mathcal{A} \quad = \quad \{\, \lambda_1\mathbf{a}_1 + \cdots + \lambda_n\mathbf{a}_n \; : \; \lambda_1, \ldots, \lambda_n \in \mathbf{N} \,\}.$$

The map π lifts to a homomorphism of semigroup algebras:

$$\hat{\pi} : k[\mathbf{x}] \rightarrow k[\mathbf{t}^{\pm 1}], \; x_i \mapsto \mathbf{t}^{\mathbf{a}_i}. \tag{4.2}$$

The kernel of $\hat{\pi}$ is denoted $I_{\mathcal{A}}$ and called the *toric ideal* of $\mathcal{A}$. Clearly, $I_{\mathcal{A}}$ is a prime ideal, and hence its affine variety $\mathcal{V}(I_{\mathcal{A}})$ of zeros in k^n is irreducible. It is the Zariski closure of the set of points $(\mathbf{t}^{\mathbf{a}_1}, \ldots, \mathbf{t}^{\mathbf{a}_n})$, where $\mathbf{t} \in (k^*)^d$. Here k^* denotes $k \setminus \{0\}$. The multiplicative group $(k^*)^d$ is called the *(d-dimensional algebraic) torus.* A variety of the form $\mathcal{V}(I_{\mathcal{A}})$ is an *affine toric variety.* This differs from the definition of "toric variety" found in the algebraic geometry literature (cf. (Fulton 1993)) in that we do not require toric varieties to be normal. The issue of normality and relations to algebraic geometry will be discussed in Chapter 13. Our first lemma specifies an infinite generating set for the toric ideal $I_{\mathcal{A}}$.

Lemma 4.1. *The toric ideal $I_{\mathcal{A}}$ is spanned as a k-vector space by the set of binomials*

$$\{\, \mathbf{x}^{\mathbf{u}} - \mathbf{x}^{\mathbf{v}} \; : \; \mathbf{u}, \mathbf{v} \in \mathbf{N}^n \;\; \textit{with} \;\; \pi(\mathbf{u}) = \pi(\mathbf{v}) \}. \tag{4.3}$$

Proof: A binomial $\mathbf{x}^{\mathbf{u}} - \mathbf{x}^{\mathbf{v}}$ lies in $I_{\mathcal{A}}$ if and only if $\pi(\mathbf{u}) = \pi(\mathbf{v})$. It therefore suffices to show that each polynomial in $I_{\mathcal{A}}$ is a k-linear combination of these binomials. Fix a term order $\prec$ on $k[\mathbf{x}]$. Suppose $f \in I_{\mathcal{A}}$ cannot be written as a k-linear combination of binomials. We choose a polynomial f with this property such that the initial term $in_{\prec}(f) = \mathbf{x}^{\mathbf{u}}$ is minimal with respect to the term order $\prec$. When expanding $f(\mathbf{t}^{\mathbf{a}_1}, \ldots, \mathbf{t}^{\mathbf{a}_n})$ we get zero. In particular, the term $\mathbf{t}^{\pi(\mathbf{u})} = \hat{\pi}(\mathbf{x}^{\mathbf{u}})$ must cancel during this expansion. Hence there is some other monomial $\mathbf{x}^{\mathbf{v}} \prec \mathbf{x}^{\mathbf{u}}$ appearing in f such that $\pi(\mathbf{v}) = \pi(\mathbf{u})$. Also the polynomial $f' := f - \mathbf{x}^{\mathbf{u}} + \mathbf{x}^{\mathbf{v}}$ cannot be written as a k-linear combination of binomials in $I_{\mathcal{A}}$. But since $in_{\prec}(f') \prec in_{\prec}(f)$, this is a contradiction. ∎

We next compute the dimension of the toric variety $\mathcal{V}(I_\mathcal{A})$. We write $\mathbf{Z}\mathcal{A}$ for the lattice spanned by $\mathcal{A}$ and $dim(\mathcal{A})$ for the dimension of $\mathbf{Z}\mathcal{A}$.

Lemma 4.2. *The Krull dimension of the residue ring $k[\mathbf{x}]/I_\mathcal{A}$ is equal to $dim(\mathcal{A})$.*

Proof: The ring $k[\mathbf{x}]/I_\mathcal{A}$ is isomorphic to the subring $k[\mathbf{t}^{\mathbf{a}_1},\ldots,\mathbf{t}^{\mathbf{a}_n}]$ of $k[\mathbf{t}^{\pm 1}]$. The Krull dimension of this integral domain is the maximum number of algebraically independent monomials $\mathbf{t}^{a_i}$. But a set of monomials is algebraically independent if and only if their exponent vectors are linearly independent (by Lemma 4.1). ∎

Every vector $\mathbf{u} \in \mathbf{Z}^n$ can be written uniquely as $\mathbf{u} = \mathbf{u}^+ - \mathbf{u}^-$ where $\mathbf{u}^+$ and $\mathbf{u}^-$ are non-negative and have disjoint support. (More precisely, the i-th coordinate of $\mathbf{u}^+$ equals u_i if $u_i > 0$ and it equals 0 otherwise.) We write $ker(\pi)$ for the sublattice of $\mathbf{Z}^n$ consisting of all vectors $\mathbf{u}$ such that $\pi(\mathbf{u}^+) = \pi(\mathbf{u}^-)$. Lemma 4.1 can be rephrased as follows.

Corollary 4.3. $I_\mathcal{A} = \langle \mathbf{x}^{\mathbf{u}^+} - \mathbf{x}^{\mathbf{u}^-} \; : \; \mathbf{u} \in ker(\pi) \rangle$.

Corollary 4.4. *For every term order $\prec$ there is a finite set of vectors $\mathcal{G}_\prec \subset ker(\pi)$ such that the reduced Gröbner basis of $I_\mathcal{A}$ with respect to $\prec$ is equal to $\{ \mathbf{x}^{\mathbf{u}^+} - \mathbf{x}^{\mathbf{u}^-} \; : \; \mathbf{u} \in \mathcal{G}_\prec \}$.*

Proof: By the Hilbert Basis Theorem we can select a finite subset of $ker(\pi)$ such that the corresponding binomials generate $I_\mathcal{A}$. Apply the Buchberger algorithm to these binomials. The operations of reduction and forming S-pairs preserve the binomial structure. Any new polynomial occurring during this run of the Buchberger algorithm (and in particular its output) lies in $\{ \mathbf{x}^{\mathbf{u}^+} - \mathbf{x}^{\mathbf{u}^-} \; : \; \mathbf{u} \in ker(\pi) \}$ as well. ∎

The Buchberger algorithm for toric ideals is a purely combinatorial process involving lattice vectors. This is worked out in detail in (Thomas 1994). Throughout this book we frequently switch back and forth between lattice vectors $\mathbf{u}$ and their associated binomials $\mathbf{x}^{\mathbf{u}^+} - \mathbf{x}^{\mathbf{u}^-}$. In particular, we shall refer to the set of vectors $\mathcal{G}_\prec$ as the *reduced Gröbner basis of $\mathcal{A}$ with respect to $\prec$*. This set can be computed "from scratch" as follows:

Algorithm 4.5. *(Computing a first Gröbner basis of a toric ideal)*

1. Introduce $n + d + 1$ indeterminates $t_0, t_1, \ldots, t_d, x_1, x_2, \ldots, x_n$. Let $\prec$ be any elimination term order with $\{t_i\} \succ \{x_j\}$.
2. Compute the reduced Gröbner basis $\mathcal{G}$ for the ideal

$$\langle\, t_0 t_1 \cdots t_d - 1 ,\; x_1 \cdot \mathbf{t}^{\mathbf{a}_1^-} - \mathbf{t}^{\mathbf{a}_1^+} ,\; \ldots ,\; x_n \cdot \mathbf{t}^{\mathbf{a}_n^-} - \mathbf{t}^{\mathbf{a}_n^+} \,\rangle. \tag{4.4}$$

3. Output: The set $\mathcal{G} \cap k[\mathbf{x}]$ is the reduced Gröbner basis for $I_\mathcal{A}$ with respect to $\prec$.

The correctness of Algorithm 4.5 is a special case of Theorem 2 in §3.3 of (Cox, Little & O'Shea 1992). Often the given lattice points $\mathbf{a}_i$ will have non-negative coordinates. In that case the variable t_0 is unnecessary, and instead of (4.4) we can use the ideal $\langle\, x_i - \mathbf{t}^{a_i} \; : \; i = 1, \ldots, n \,\rangle$. Algorithm 4.5 is by no means best possible. Faster algorithms for finding generators and Gröbner bases of toric ideals will be given in Chapter 12.

We define the *universal Gröbner basis* $\mathcal{U}_\mathcal{A}$ to be the union of all reduced Gröbner bases $\mathcal{G}_\prec$ of the toric ideal $I_\mathcal{A}$ as $\prec$ runs over all term orders. Theorem 1.2 and Corollary 4.4 imply that $\mathcal{U}_\mathcal{A}$ is a finite set consisting of binomials. We can thus identify $\mathcal{U}_\mathcal{A}$ with a finite set of vectors in $ker(\pi)$. Our goal is to find a more precise description of the universal Gröbner basis. A binomial $\mathbf{x}^{\mathbf{u}^+} - \mathbf{x}^{\mathbf{u}^-}$ in $I_\mathcal{A}$ is called *primitive* if there exists no other binomial $\mathbf{x}^{\mathbf{v}^+} - \mathbf{x}^{\mathbf{v}^-} \in I_\mathcal{A}$ such that $\mathbf{x}^{\mathbf{v}^+}$ divides $\mathbf{x}^{\mathbf{u}^+}$ and $\mathbf{x}^{\mathbf{v}^-}$ divides $\mathbf{x}^{\mathbf{u}^-}$.

Lemma 4.6. *Every binomial $\mathbf{x}^{\mathbf{u}^+} - \mathbf{x}^{\mathbf{u}^-}$ in the universal Gröbner basis $\mathcal{U}_\mathcal{A}$ is primitive.*

Proof: Let $\mathbf{x}^{\mathbf{u}^+} - \mathbf{x}^{\mathbf{u}^-}$ be any binomial in the reduced Gröbner basis $\mathcal{G}_\prec$, and let $\mathbf{u}^+ \succ \mathbf{u}^-$. Then $\mathbf{x}^{\mathbf{u}^+}$ is a minimal generator of $in_\prec(I_\mathcal{A})$ and $\mathbf{x}^{\mathbf{u}^-}$ is a standard monomial. Suppose that $\mathbf{x}^{\mathbf{u}^+} - \mathbf{x}^{\mathbf{u}^-}$ is not primitive. Choose $\mathbf{v} \in ker(\pi)$ with $\mathbf{v} \neq \mathbf{u}$ such that $\mathbf{x}^{\mathbf{v}^+}$ divides $\mathbf{x}^{\mathbf{u}^+}$ and $\mathbf{x}^{\mathbf{v}^-}$ divides $\mathbf{x}^{\mathbf{u}^-}$. If $\mathbf{v}^+ \succ \mathbf{v}^-$ then $\mathbf{x}^{\mathbf{u}^+}$ is not a minimal generator of $in_\prec(I_\mathcal{A})$, a contradiction. If $\mathbf{v}^+ \prec \mathbf{v}^-$ then $\mathbf{x}^{\mathbf{u}^-}$ is not standard, a contradiction. ∎

The converse to Lemma 4.6 is not true. There may be primitive binomials which do not appear in $\mathcal{U}_\mathcal{A}$. For instance, take $n = 3, d = 1, \mathcal{A} = \{1, 2, 4\}$. Then $x_1^2x_2 - x_3$ is a primitive binomial in $I_\mathcal{A}$ but it does not appear in $\mathcal{U}_\mathcal{A} = \{x_1^2 - x_2,\, x_1^4 - x_3,\, x_2^2 - x_3\}$. In general, however, the set of primitive binomials is a pretty good approximation to the universal Gröbner basis. We shall prove the following degree bound.

Theorem 4.7. *Let $dim(\mathcal{A}) = d$ and $D(\mathcal{A}) := max\,\{|det(\mathbf{a}_{i_1}, \ldots, \mathbf{a}_{i_d})| \,:\, 1 \leq i_1 < \cdots < i_d \leq n\}$. The total degree of any primitive binomial in $I_\mathcal{A}$ is less than $(d+1)(n-d)D(\mathcal{A})$.*

We identify $\mathcal{A}$ with the $d \times n$-matrix $(\mathbf{a}_1, \ldots, \mathbf{a}_n)$. The hypothesis in Theorem 4.7 is that the matrix $\mathcal{A}$ has maximal rank d. If this hypothesis does not hold, then one can delete some rows from $\mathcal{A} = (a_{ij})$ until it holds. For our proof of Theorem 4.7 we need to introduce a distinguished subset of primitive elements. A non-zero vector $\mathbf{u}$ in $ker(\pi)$ is called a *circuit* if its support $supp(\mathbf{u})$ is minimal with respect to inclusion and the coordinates of $\mathbf{u}$ are relatively prime. (This definition is consistent with the one given in Example 1.5.) Equivalently, a circuit is an irreducible binomial $\mathbf{x}^{\mathbf{u}^+} - \mathbf{x}^{\mathbf{u}^-}$ in $I_\mathcal{A}$ which has minimal support. It is easy to see that every circuit is primitive.

Lemma 4.8. *If $\mathbf{u}$ is a circuit in $ker(\pi)$ then $supp(\mathbf{u})$ has at most $d+1$ elements.*

Proof: Suppose $\mathbf{u} \in ker(\pi)$ has $r \geq d+2$ non-zero coordinates. Let B be the $d \times r$-submatrix of $\mathcal{A}$ given by these column indices. The kernel of B is at least 2-dimensional and hence contains a non-zero vector $\mathbf{v}'$ with at least one zero coordinate. Extend $\mathbf{v}'$ to a non-zero vector $\mathbf{v} \in ker(\pi)$ by placing zeros in the other $n - r$ coordinates. Then $supp(\mathbf{v})$ is a proper subset of $supp(\mathbf{u})$, which shows that $\mathbf{u}$ is not a circuit. ∎

Lemma 4.9. *If $\mathbf{u} = (u_1, \ldots, u_n)$ is a circuit in $ker(\pi)$ then $|u_i| \leq D(\mathcal{A})$ for all i.*

Proof: Let $supp(\mathbf{u}) = \{i_1, \ldots, i_r\}$. The $d \times r$-matrix $(\mathbf{a}_{i_1}, \ldots, \mathbf{a}_{i_r})$ has rank $r - 1$, by the same argument as in the proof of Lemma 4.8. Since $\mathcal{A}$ has rank d, we can find column vectors $\mathbf{a}_{i_{r+1}}, \ldots, \mathbf{a}_{i_{d+1}}$ such that the $d \times (d+1)$-matrix

$B = (\mathbf{a}_{i_1}, \ldots, \mathbf{a}_{i_r}, \mathbf{a}_{i_{r+1}}, \ldots, \mathbf{a}_{i_{d+1}})$ has rank d. Let $\mathbf{e}_i$ denote the i-th unit vector in $\mathbf{Z}^n$. Using Cramer's rule of elementary linear algebra, we see that the kernel of B is spanned by the vector

$$\sum_{j=1}^{d+1} (-1)^j \cdot det(\mathbf{a}_{i_1}, \ldots, \mathbf{a}_{i_{j-1}}, \mathbf{a}_{i_{j+1}}, \ldots, \mathbf{a}_{i_{d+1}}) \cdot \mathbf{e}_{i_j}. \tag{4.5}$$

The restriction of $\mathbf{u}$ to $\{i_1, \ldots, i_{d+1}\}$ lies in the kernel of B and hence is a rational multiple of (4.5). Now, since (4.5) is an integer vector, and since $\mathbf{u}$ is a circuit, we conclude that (4.5) is an integer multiple of $\mathbf{u}$. This proves the claim. ∎

Let $\mathbf{u}, \mathbf{v} \in \mathbf{Z}^n$. We say that $\mathbf{u}$ is *conformal to* $\mathbf{v}$ if $supp(\mathbf{u}^+) \subset supp(\mathbf{v}^+)$ and $supp(\mathbf{u}^-) \subset supp(\mathbf{v}^-)$.

Lemma 4.10. *Every vector* $\mathbf{v}$ *in* $ker(\pi)$ *can be written as a non-negative rational linear combination of* $n-d$ *circuits each of which is conformal to* $\mathbf{v}$.

Proof: We fix d and proceed by induction on n. If $n \leq d+1$ then the assertion is trivial. Thus suppose $n \geq d+2$, and let $\mathbf{v}$ be a non-circuit in $ker(\pi)$. We may also assume that $supp(\mathbf{v}) = \{1, \ldots, n\}$, because otherwise we could delete redundant columns of $\mathcal{A}$ and use the induction hypothesis to write $\mathbf{v}$ as a conformal rational linear combination of $card(supp(\mathbf{v})) - d \leq n-d$ circuits. Let $\mathbf{u} = (u_1, \ldots, u_n)$ be any circuit such that $u_1 v_1 > 0$. Among all positive coordinate ratios v_i/u_i let λ denote the minimum. Then $\mathbf{v} - \lambda\mathbf{u}$ is conformal to $\mathbf{v}$ and has zero i-th coordinate. By the induction hypothesis, the vector $\mathbf{v} - \lambda\mathbf{u}$ can be written as a conformal rational linear combination of $n-d-1$ circuits. The identity $\mathbf{v} = \lambda\mathbf{u} + (\mathbf{v} - \lambda\mathbf{u})$ now completes the proof. ∎

Proof of Theorem 4.7. Let $\mathbf{v}$ be a primitive vector in $ker(\pi)$. If $\mathbf{v}$ is a circuit, then we are done by Lemma 4.9. If not, then we apply Lemma 4.10 to find circuits $\mathbf{u}_1, \ldots, \mathbf{u}_{n-d}$, each conformal to $\mathbf{v}$, and non-negative rationals $\lambda_1, \ldots, \lambda_{n-d}$ such that

$$\mathbf{v} \;=\; \lambda_1 \mathbf{u}_1 \;+\; \cdots \;+\; \lambda_{n-d}\mathbf{u}_{n-d}. \tag{4.6}$$

The fact that each $\mathbf{u}_i$ is conformal to $\mathbf{v}$ means that $\mathbf{v}^+ \;=\; \lambda_1 \mathbf{u}_1^+ + \cdots + \lambda_{n-d}\mathbf{u}_{n-d}^+$ and $\mathbf{v}^- \;=\; \lambda_1 \mathbf{u}_1^- + \cdots + \lambda_{n-d}\mathbf{u}_{n-d}^-$. This implies that each λ_i is less than 1, because otherwise $\mathbf{v}$ would not be primitive.

We may assume that the total degree of the binomial $\mathbf{x}^{\mathbf{v}^+} - \mathbf{x}^{\mathbf{v}^-}$ equals $||\mathbf{v}^+||_1$, the coordinate sum of the positive part $\mathbf{v}^+$. Applying the norm inequality for the 1-norm to the positive part of (4.6), we get

$$\begin{aligned} ||\mathbf{v}^+||_1 \;\; &\leq \;\; \sum_{i=1}^{n-d} \lambda_j \cdot ||\mathbf{u}_j^+||_1 \;\; < \\ (n-d) \cdot max\{||\mathbf{u}_j^+||_1 \,:\, j = 1, \ldots, n-d\} \;\; &\leq \;\; (n-d)\cdot(d+1)\cdot D(\mathcal{A}). \end{aligned} \tag{4.7}$$

In the last step we used the inequality $||\mathbf{u}_j^+||_1 \;\leq\; (d+1)\cdot D(\mathcal{A})$, which follows directly from Lemma 4.8 and Lemma 4.9. ∎

The set of circuits in $I_{\mathcal{A}}$ is denoted by $\mathcal{C}_{\mathcal{A}}$. One method for computing $\mathcal{C}_{\mathcal{A}}$ is to evaluate Cramer's determinantal formula (4.5) for all $(d+1)$-subsets $\{i_1,\ldots,i_{d+1}\}$ of $\{1,\ldots,n\}$. Obviously, such a computation must be organized in a clever manner to be of practical use for larger values of n and d. We call the set of primitive binomials the *Graver basis* of $\mathcal{A}$ and denote it by $Gr_{\mathcal{A}}$. This name is in reference to the work of Jack Graver on integer programming in (Graver 1975). The connection to integer programming is discussed in Chapter 5. In Chapter 7 we present algorithms for computing the Graver basis $Gr_{\mathcal{A}}$ and the universal Gröbner basis $\mathcal{U}_{\mathcal{A}}$.

Proposition 4.11. *For every finite set* $\mathcal{A} \subset \mathbf{Z}^d$ *we have* $\mathcal{C}_{\mathcal{A}} \subseteq \mathcal{U}_{\mathcal{A}} \subseteq Gr_{\mathcal{A}}$.

Proof: The inclusion $\mathcal{U}_{\mathcal{A}} \subseteq Gr_{\mathcal{A}}$ was shown in Lemma 4.6. It remains to be shown that each circuit lies in some reduced Gröbner basis. Let $\mathbf{u} \in ker(\pi)$ be a circuit. Fix an elimination term order $\prec$ such that $\{x_i : i \notin supp(\mathbf{u})\} \succ \{x_j : j \in supp(\mathbf{u})\}$ and $x^{\mathbf{u}^+} \succ x^{\mathbf{u}^-}$. We claim that $\mathbf{x}^{\mathbf{u}^+} - \mathbf{x}^{\mathbf{u}^-}$ appears in the reduced Gröbner basis $\mathcal{G}_\prec$ of $I_{\mathcal{A}}$. Suppose not. Then there exists $\mathbf{v} \in ker(\pi)\backslash\{0,\mathbf{u}\}$ such that $x^{\mathbf{v}^+} \succ x^{\mathbf{v}^-}$ and $\mathbf{x}^{\mathbf{v}^+}$ divides $\mathbf{x}^{\mathbf{u}^+}$. The choice of term order and the inclusion $supp(\mathbf{v}^+) \subseteq supp(\mathbf{u})$ implies $supp(\mathbf{v}^-) \subseteq supp(\mathbf{u})$, and hence $supp(\mathbf{v}) \subseteq supp(\mathbf{u})$. Since $\mathbf{u}$ is a circuit, this implies that $\mathbf{v}$ is an integer multiple of $\mathbf{u}$. In view of $\mathbf{x}^{\mathbf{v}^+}$ dividing $\mathbf{x}^{\mathbf{u}^+}$, this is only possible if $\mathbf{u} = \mathbf{v}$. ∎

Example 4.12. *(The inclusions in Proposition 4.11 may or may not be strict)*
If $n = 3, d = 1$ and the integers in $\mathcal{A} = \{i,j,k\} \subset \mathbf{N}$ are pairwise relatively prime, then $\mathcal{C}_{\mathcal{A}} = \{x_1^j - x_2^i, x_1^k - x_3^i, x_2^k - x_3^j\}$. We consider the following three such cases:

- If $\mathcal{A} = \{1,2,3\}$ then $\mathcal{U}_{\mathcal{A}} = Gr_{\mathcal{A}} = \mathcal{C}_{\mathcal{A}} \cup \{x_3 - x_1x_2, x_1x_3 - x_2^2\}$.
- If $\mathcal{A} = \{1,2,4\}$ then $\mathcal{C}_{\mathcal{A}} = \mathcal{U}_{\mathcal{A}}$ and $Gr_{\mathcal{A}} \setminus \mathcal{U}_{\mathcal{A}} = \{x_3 - x_1^2x_2\}$.
- If $\mathcal{A} = \{1,2,5\}$ then $\mathcal{U}_{\mathcal{A}} \setminus \mathcal{C}_{\mathcal{A}} = \{x_3 - x_1x_2^2, x_1x_3 - x_2^3\}$ and $Gr_{\mathcal{A}} \setminus \mathcal{U}_{\mathcal{A}} = \{x_3 - x_1^3x_2\}$.

But there are also plenty of sets $\mathcal{A}$ for which $\mathcal{C}_{\mathcal{A}} = \mathcal{U}_{\mathcal{A}} = Gr_{\mathcal{A}}$ holds. This is the case for

$$\begin{aligned}\mathcal{A} \;:=\; \big\{&(1,0,1,0,0),(1,0,0,1,0),(1,0,0,0,1),\\ &(0,1,1,0,0),(0,1,0,1,0),(0,1,0,0,1)\big\}.\end{aligned}$$

Here $I_{\mathcal{A}}$ equals the ideal of 2×2-minors of a 2×3-matrix of indeterminates (Example 1.4). The three 2×2-minors are the circuits of $\mathcal{A}$ and they are also the Graver basis of $\mathcal{A}$. ∎

We note the following general facts concerning elimination ideals of toric ideals.

Proposition 4.13. *Let* $\mathcal{B}$ *be any subset of* $\mathcal{A}$ *and let* $k[\mathcal{B}] := k[x_i : \mathbf{a}_i \in \mathcal{B}]$. *Then*

(a) *the toric ideal of* $\mathcal{B}$ *is* $I_{\mathcal{B}} = I_{\mathcal{A}} \cap k[\mathcal{B}]$;
(b) *the circuits of* $\mathcal{B}$ *are* $\mathcal{C}_{\mathcal{B}} = \mathcal{C}_{\mathcal{A}} \cap k[\mathcal{B}]$;
(c) *the universal Gröbner basis of* $\mathcal{B}$ *is* $\mathcal{U}_{\mathcal{B}} = \mathcal{U}_{\mathcal{A}} \cap k[\mathcal{B}]$;
(d) *the Graver basis of* $\mathcal{B}$ *is* $Gr_{\mathcal{B}} = Gr_{\mathcal{A}} \cap k[\mathcal{B}]$.

Proof: Left to the reader; see Exercise (3) below. ∎

A typical feature of many sets $\mathcal{A}$ arising in practise is that their toric ideal $I_\mathcal{A}$ is homogeneous. Here "homogeneous" refers to the total degree grading given by $deg(x_1) = \cdots = deg(x_n) = 1$. If $I_\mathcal{A}$ is homogeneous then its zero set $\mathcal{V}(I_\mathcal{A})$ in projective space P^{n-1} is a *projective toric variety*. Its dimension equals $dim(\mathcal{A})-1$ (by Lemma 4.2). The second most important invariant of a projective variety (besides its dimension) is its *degree*. In the remainder of this chapter we determine the degree of the projective toric variety $\mathcal{V}(I_\mathcal{A})$.

Lemma 4.14. *The ideal $I_\mathcal{A}$ is homogeneous if and only if there exists a vector $\omega \in \mathbf{Q}^d$ such that $\mathbf{a}_i \cdot \omega = 1$ for $i = 1, \ldots, n$.*

Proof: A binomial $\mathbf{x}^{\mathbf{u}^+} - \mathbf{x}^{\mathbf{u}^-}$ is homogeneous if and only if the vector $\mathbf{u} = \mathbf{u}^+ - \mathbf{u}^-$ has coordinate sum zero. In view of Corollary 4.3, $I_\mathcal{A}$ is homogeneous if and only if all vectors $\mathbf{u} \in ker(\pi)$ have zero coordinate sum. This holds if and only if $(1,1,\ldots,1)$ lies in the subspace $ker(\pi)^\perp = image(\pi^T) = image\big(\omega \mapsto (\mathbf{a}_1 \cdot \omega, \ldots, \mathbf{a}_n \cdot \omega)\big)$ of $\mathbf{R}^n$. ∎

We remark that in the homogeneous case the degree bound of Theorem 4.7 can easily be improved by a factor of 2.

Corollary 4.15. *Let $\mathcal{A}$ be as in Lemma 4.14. Then the total degree of any primitive binomial in the homogeneous toric ideal $I_\mathcal{A}$ is less than $\frac{1}{2}(d+1)(n-d)D(\mathcal{A})$.*

Proof: In the homogeneous case, each circuit $\mathbf{u}_j$ satisfies $||\mathbf{u}_j^+||_1 = ||\mathbf{u}_j^-||_1 \leq \frac{1}{2}(d+1)D(\mathcal{A})$, by Lemmas 4.8 and 4.9. Now use the inequality (4.7) as in the proof of Theorem 4.7. ∎

It is an interesting question whether the maximum degree of any Graver basis element is always attained by a circuit. In other words, is

$$max\, deg(Gr_\mathcal{A}) \quad = \quad max\, deg(\mathcal{C}_\mathcal{A}) \qquad \text{for all sets } \mathcal{A}\text{ ?}$$

The answer is "no". A counterexample was contructed after this book had been submitted to the publisher. It will appear in a future publication.

Lemma 4.14 states in geometric terms that $I_\mathcal{A}$ is homogeneous if and only if the points of $\mathcal{A}$ lie on a common affine hyperplane in $\mathbf{R}^d$. Assuming that this is the case, we introduce the polytope $Q = conv(\mathcal{A})$. The *normalized Ehrhart polynomial* of Q is the numerical function

$$E_Q : \mathbf{N} \to \mathbf{N}, \quad r \mapsto card(\mathbf{Z}\mathcal{A} \cap r \cdot Q). \tag{4.8}$$

It is known that $E_Q(t)$ is a polynomial of degree $q := dim(Q) \leq d-1$, see e.g. (Stanley 1986). Hence we can write $E_Q(r) = \sum_{i=0}^{q} c_i/i! \cdot r^i$. The leading coefficient c_q is denoted $Vol(Q)$ and called the *normalized volume* of Q. If $q = d-1$ then $Vol(Q)$ equals the usual Euclidean volume of Q times $q!$ times the order of the finite abelian group $\mathbf{Z}^d/\mathbf{Z}\mathcal{A}$.

Theorem 4.16. *The degree of the projective toric variety defined by $I_\mathcal{A}$ equals the normalized volume $Vol(Q)$ of the polytope $Q = conv(\mathcal{A})$.*

Proof: The degree of the projective toric variety $\mathcal{V}(I_\mathcal{A})$ equals $q!$ times the leading coefficient of the Hilbert polynomial $H_\mathcal{A}(r)$ of $k[\mathbf{x}]/I_\mathcal{A}$. The Hilbert polynomial is defined as follows: $H_\mathcal{A}(r)$ is the k-dimension of the r-th graded component of

$k[\mathbf{x}]/I_{\mathcal{A}} = k[\mathbf{t}^{\mathbf{a}_1}, \ldots, \mathbf{t}^{\mathbf{a}_n}]$ for $r \gg 0$. A basis for this vector space is given by the set of all monomials $\mathbf{t}^{\mathbf{b}}$ such that $\mathbf{b}$ lies in the semigroup spanned by $\mathcal{A}$ and $\mathbf{b} \cdot \omega = r$, where $\omega \in \mathbf{Q}^d$ is as in Lemma 4.14. In other words, $H_{\mathcal{A}}(r) = card(\mathbf{N}\mathcal{A} \cap r \cdot Q)$. This implies the inequality $H_{\mathcal{A}}(r) \leq E_Q(r)$.

Consider the following finite set

$$C \quad := \quad \mathbf{Z}\mathcal{A} \cap \bigl\{ \sum_{i=1}^{n} \lambda_i \mathbf{a}_i \; : \; 0 \leq \lambda_i \leq 1 \text{ for } i = 1, \ldots, n \bigr\}.$$

Fix an integer $R > 0$ such that each $\mathbf{b} \in C$ can be written as $\mathbf{b} = \pi(\mathbf{u})$ for some $\mathbf{u} = (u_1, \ldots, u_n) \in \mathbf{Z}^n$ with $u_i > -R$ for all i. For any integer $r > nR$ consider the map

$$\mathbf{Z}\mathcal{A} \cap (r - nR) \cdot Q \;\rightarrow\; \mathbf{N}\mathcal{A} \cap r \cdot Q, \quad \mathbf{b} \mapsto \mathbf{b} + R \cdot (\mathbf{a}_1 + \cdots + \mathbf{a}_n) \tag{4.9}$$

To show that this map is well-defined, we must express $\mathbf{b} + R \cdot (\mathbf{a}_1 + \cdots + \mathbf{a}_n)$ as a *non-negative* integer linear combination of $\mathcal{A}$. This can be done as follows: first write $\mathbf{b}$ as a non-negative rational linear combination of $\mathcal{A}$. If any of the rational coefficients λ_i is ≥ 1, then we subtract the corresponding $\mathbf{a}_i$ from the representation (this does not harm the conclusion). Repeating this process, we eventually reduce to the case $\mathbf{b} \in C$. Now write $\mathbf{b} = \sum u_i \mathbf{a}_i$ with integers $u_i > -R$. Adding R times the sum $\mathbf{a}_1 + \cdots + \mathbf{a}_n$ to this representation, we obtain the desired conclusion.

Clearly, the map (4.9) is injective, and therefore

$$E_Q(r - nR) \quad \leq \quad H_{\mathcal{A}}(r) \quad \leq \quad E_Q(r). \tag{4.10}$$

This shows that the polynomials E_Q and $H_{\mathcal{A}}$ have the same degree and the same leading coefficient, and the proof is complete. ∎

Example 4.17. *(The Segre variety $P^1 \times P^1 \times P^1$ in P^7)*
Let $\mathcal{A}$ be the set of vertices of a regular 3-dimensional cube, given in homogeneous coordinates. For instance, $\mathcal{A} = \bigl\{(1,0,1,0,1,0), (1,0,0,1,1,0), (1,0,1,0,0,1), (1,0,0,1,0,1), (0,1,1,0,1,0), (0,1,0,1,1,0), (0,1,1,0,0,1), (0,1,0,1,0,1)\bigr\}$. The normalized volume of the regular cube $Q = conv(\mathcal{A})$ is six. Hence $\mathcal{V}(I_{\mathcal{A}})$ is a projective toric variety of dimension 3 and degree 6. It is the Segre embedding of $P^1 \times P^1 \times P^1$ in P^7. ∎

Exercises:

(1) Find a generating set for the ideal $I_{\mathcal{A}}$ of the Segre-threefold in Example 4.17. Also compute the circuits $\mathcal{C}_{\mathcal{A}}$, the universal Gröbner basis $\mathcal{U}_{\mathcal{A}}$, and the Graver basis $Gr_{\mathcal{A}}$.

(2) For fixed integers d and r consider the set $\mathcal{A} = \bigl\{ (i_1, i_2, \ldots, i_d) \in \mathbf{N}^d : i_1 + i_2 + \cdots + i_d = r \bigr\}$. Show that $I_{\mathcal{A}}$ has a reduced Gröbner basis consisting of quadratic binomials.

(3) Prove Proposition 4.13.

(4) For the set $\mathcal{A} = \{(1,0), (1,1), (1,5)\}$ (graded with $\omega = (1,0)$) compute the Hilbert polynomial $H_{\mathcal{A}}$ and the normalized Ehrhart polynomial E_Q.

(5) Does the hypothesis $\mathcal{A} = \mathbf{Z}\mathcal{A} \cap Q$ imply the conclusion $H_{\mathcal{A}} = E_Q$?

(6) Let $\mathcal{A}$ be the set of all $m \times m$-permutation matrices. (Here $n = m!$ and $d = m^2$).
   (a) For $m = 3$ show that $I_{\mathcal{A}}$ is a principal ideal.
   (b) For $m = 4$ compute any Gröbner basis for $I_{\mathcal{A}}$.
   (c) For $m = 5$ determine the degree of the projective toric variety $\mathcal{V}(I_{\mathcal{A}})$.
   (d) For $m = 6$ give an upper bound on the degree and cardinality of the universal Gröbner basis $\mathcal{U}_{\mathcal{A}}$.

(7) Prove the following sufficient condition for a toric ideal to be a complete intersection: If the origin lies in the interior of the convex hull of $\mathcal{A}$ in $\mathbf{R}^d$, then $I_{\mathcal{A}}$ is generated by $n - d$ binomials. Does there always exist a Gröbner basis consisting of $n - d$ binomials ?

(8) Let $\mathcal{A} = \{\mathbf{e}_i - \mathbf{e}_j \,:\, 1 \le i < j \le n\}$. Show that $\mathcal{C}_{\mathcal{A}} = \mathcal{U}_{\mathcal{A}} = Gr_{\mathcal{A}}$. Identify this set of binomials with the circuits in a directed complete graph on n nodes.

Notes:

Semigroup algebras and their presentation ideals (here called "toric ideals") have been studied by many researchers. The emphasis of most publications lies on commutative algebra issues, such as Cohen-Macaulayness and local cohomology, with the strongest results being typically available for monomial curves ($d = 1$). An early reference is (Herzog 1970), and three more recent ones are (Trung & Hoa 1986), (Bresinsky 1988), and (Campillo & Pison 1993). The single-exponential degree bound for Gröbner bases of toric ideals in Theorem 4.7 appeared in (Sturmfels 1991). The relation between volume and degree in Theorem 4.16 is a staple of toric geometry. For more detailed discussions in the contexts of commutative algebra and algebraic geometry see Theorem 6.3.12 in (Bruns & Herzog 1993) and §5.3 in (Fulton 1993) respectively.

CHAPTER 5

Enumeration, Sampling and Integer Programming

In this chapter we present applications of toric ideals. We shall assume for simplicity that $\mathcal{A} = \{\mathbf{a}_1, \ldots, \mathbf{a}_n\}$ lies in $\mathbf{N}^d \setminus \{0\}$, i.e., each $\mathbf{a}_i$ is non-zero and non-negative. This assumption implies the following property for the map π in (4.1): for each $\mathbf{b} \in \mathbf{N}^d$ the set $\pi^{-1}(\mathbf{b}) = \{\mathbf{u} \in \mathbf{N}^n : \pi(\mathbf{u}) = \mathbf{b}\}$ is finite. We call $\pi^{-1}(\mathbf{b})$ the *fiber* of $\mathcal{A}$ over $\mathbf{b}$.

There are three natural families of problems associated with these fibers:

- **Enumeration:** *Determine the cardinality of $\pi^{-1}(\mathbf{b})$. If this number is "not too big", then list all elements in the fiber $\pi^{-1}(\mathbf{b})$ explicitly.*

If complete enumeration is infeasible, then the following questions are of interest.

- **Sampling:** *Choose a point at random from $\pi^{-1}(\mathbf{b})$. For our purposes it suffices to assume that "at random" refers to the uniform distribution on $\pi^{-1}(\mathbf{b})$.*
- **Integer Programming:** *Given any "cost vector" $\omega \in \mathbf{R}^n$, find a point $\mathbf{u}$ in $\pi^{-1}(\mathbf{b})$ which minimizes the value of the linear functional $\mathbf{u} \mapsto \mathbf{u} \cdot \omega$.*

Our objective is to apply toric ideals to model and solve these problems. We start by illustrating the basic ideas with a simple but important example.

Example 5.1. *(Contingency tables, transportation problems, and 2×2-minors)* Fix positive integers s and t. Let $\mathbf{e}_1, \ldots, \mathbf{e}_s$ be the unit vectors in $\mathbf{N}^s$ and $\mathbf{e}'_1, \ldots, \mathbf{e}'_t$ the unit vectors in $\mathbf{N}^t$. Let $d = s+t$ and $n = s \cdot t$. We identify $\mathbf{N}^d$ with $\mathbf{N}^s \oplus \mathbf{N}^t$, and we identify $\mathbf{N}^n$ with the set $\mathbf{N}^{s \times t}$ of non-negative integer $s \times t$-matrices $\mathbf{u} = (u_{ij})$. We define $\mathcal{A} = \{\, \mathbf{e}_i \oplus \mathbf{e}'_j \, : \, i = 1, \ldots, s, \, j = 1, \ldots, t \,\} \subset \mathbf{N}^d$. The map π computes the row and column sums of a given matrix:

$$\pi : \mathbf{N}^{s \times t} \rightarrow \mathbf{N}^{s+t}, \quad \mathbf{u} \mapsto \Big(\sum_{j=1}^{t} u_{1j}, \ldots, \sum_{j=1}^{t} u_{sj} \, ; \, \sum_{i=1}^{s} u_{i1}, \ldots, \sum_{i=1}^{s} u_{it}\Big) \tag{5.1}$$

If $\mathbf{r} = (r_1, \ldots, r_s) \in \mathbf{N}^s$ and $\mathbf{c} = (c_1, \ldots, c_t) \in \mathbf{N}^t$, then the fiber $\pi^{-1}(\mathbf{r}; \mathbf{c})$ consists of all non-negative integer $s \times t$-matrices with row sums $\mathbf{r}$ and column sums $\mathbf{c}$.

Here is an example of a very small fiber of 3 × 3-matrices ($s = t = 3, n = 9, d = 6$):

$$\pi^{-1}(1,1,2;1,1,2) \quad = \quad \left\{ \begin{pmatrix} 1 & 0 & 0 \\ 0 & 1 & 0 \\ 0 & 0 & 2 \end{pmatrix}, \begin{pmatrix} 1 & 0 & 0 \\ 0 & 0 & 1 \\ 0 & 1 & 1 \end{pmatrix}, \begin{pmatrix} 0 & 1 & 0 \\ 1 & 0 & 0 \\ 0 & 0 & 2 \end{pmatrix}, \right. \\ \left. \begin{pmatrix} 0 & 1 & 0 \\ 0 & 0 & 1 \\ 1 & 0 & 1 \end{pmatrix}, \begin{pmatrix} 0 & 0 & 1 \\ 1 & 0 & 0 \\ 0 & 1 & 1 \end{pmatrix}, \begin{pmatrix} 0 & 0 & 1 \\ 0 & 1 & 0 \\ 1 & 0 & 1 \end{pmatrix}, \begin{pmatrix} 0 & 0 & 1 \\ 0 & 0 & 1 \\ 1 & 1 & 0 \end{pmatrix} \right\}. \tag{5.2}$$

We now present somewhat more realistic instances for the three problems above.

- **Enumeration:** Suppose we wish to find all non-negative integer 3×3-matrices having row sums (42, 42, 5) and column sums (28, 30, 31). A typical such matrix looks like

$$\mathbf{u} \quad = \quad \begin{pmatrix} 20 & 13 & 9 \\ 6 & 16 & 20 \\ 2 & 1 & 2 \end{pmatrix}. \tag{5.3}$$

  It can be shown that $\pi^{-1}(42,42,5;28,30,31)$ has precisely 13,132 elements. The enumeration problem is to efficiently generate these 13,132 matrices.

- **Sampling:** The following 4 × 4-matrix was used in (Diaconis & Sturmfels 1993) to illustrate the concept of *contingency tables* in statistics. This data set classifies 592 people according to their eye color and their hair color.

Eye Color			*Hair Color*			
	Black	Brunette	Red	Blonde	Total	
Brown	68	119	26	7	220	
Blue	20	84	17	94	215	(5.4)
Hazel	15	54	14	10	93	
Green	5	29	14	16	64	
Total	108	286	71	127	592	

  A natural question to ask about these data is whether eye color and hair color are correlated ? One approach to answering this is to compare certain features (e.g. the χ^2-statistic) of the table (5.4) with that of a comparison table selected at random among all tables with the same marginal distribution. In our notation above, the set of tables with the same marginal distribution is $\pi^{-1}(220,215,93,64;108,286,71,127)$. The sampling problem is to select a random element from this fiber. Complete enumeration is infeasible in this instance. In fact, it is a non-trivial problem to even count fibers like these. Using the methods in (Mount 1995), it can be shown that

$$card\big(\pi^{-1}(220,215,93,64;108,286,71,127)\big) \quad = \quad 1,225,914,276,768,514.$$

- **Integer Programming:** The integer programming problem associated with the set $\mathcal{A} = \{\mathbf{e}_i \oplus \mathbf{e}'_j\}$ is called the *transportation problem.* Consider $s = 4$ factories F_1, F_2, F_3 and F_4 which produce a respective supply of 220, 215, 93 and 64 units of an indivisible good. Consider also $t = 4$ stores S_1, S_2, S_3 and S_4 which have respective demands of $108, 286, 71$ and 127 units. There is a (non-negative real) cost ω_{ij} associated with transporting one unit from factory F_i to store S_j. The possible transportation plans for shipping all 592 units from the factories to the stores are precisely the elements of $\pi^{-1}(220, 215, 93, 64; 108, 286, 71, 127)$. The *transportation problem* is to find a matrix $\mathbf{u} = (u_{ij})$ in that fiber which minimizes the total cost $\sum_{1\leq i,j\leq 4} u_{ij}\cdot\omega_{ij}$. Here is an example of a cost matrix and corresponding optimal transportation plan:

$$\text{If } \omega = \begin{pmatrix} 1 & 1 & 1 & 1 \\ 4 & 3 & 2 & 1 \\ 7 & 5 & 3 & 1 \\ 10 & 7 & 4 & 1 \end{pmatrix}, \text{ then } \mathbf{u} = \begin{pmatrix} 108 & 112 & 0 & 0 \\ 0 & 174 & 41 & 0 \\ 0 & 0 & 30 & 63 \\ 0 & 0 & 0 & 64 \end{pmatrix} \text{ is optimal.} \tag{5.5}$$

Returning from Example 5.1 to the general case, we shall first discuss the sampling problem. The basic idea is to do a *random walk* on the fiber $\pi^{-1}(\mathbf{b})$. Let $\mathcal{F}$ be any finite subset of $ker(\pi)$. We define a graph denoted $\pi^{-1}(\mathbf{b})_{\mathcal{F}}$ as follows. The nodes of this graph are the elements in $\pi^{-1}(\mathbf{b})$, and two nodes $\mathbf{u}$ and $\mathbf{u}'$ are connected by an edge if $\mathbf{u}-\mathbf{u}' \in \mathcal{F}$ or $\mathbf{u}'-\mathbf{u} \in \mathcal{F}$. The graph $\pi^{-1}(\mathbf{b})_{\mathcal{F}}$ may be connected, or it may be disconnected if $\mathcal{F}$ is chosen too small. If it is connected, then the following simple random walk defines a Markov chain which converges to the uniform distribution on $\pi^{-1}(\mathbf{b})$.

Algorithm 5.2. *(Random walk on a fiber)*
`Input:` A finite set $\mathcal{F} \subset ker(\pi)$ of "moves". An initial point $\mathbf{u}^{(0)}$ in a fiber $\pi^{-1}(\mathbf{b})$.
`Output:` A "random" point $\mathbf{u}$ in $\pi^{-1}(\mathbf{b})$, provided the graph $\pi^{-1}(\mathbf{b})_{\mathcal{F}}$ is connected.

1. Let $\mathbf{u} := \mathbf{u}^{(0)}$.
2. While (some termination condition is not yet satisfied) do
   2.1. select $\mathbf{v}$ at random from the uniform distribution on $\mathcal{F} \cup -\mathcal{F}$.
   2.2. if $\mathbf{u}+\mathbf{v}$ is non-negative then replace $\mathbf{u}$ by $\mathbf{u}+\mathbf{v}$.
   (if $\mathbf{u}+\mathbf{v}$ has a negative coordinate, the walk stays at $\mathbf{u}$.)

We here ignore the question of running time until stationarity and what the "termination condition" should be. The main point for us is how to find a finite set of moves $\mathcal{F}$ which is guaranteed to connect all fibers $\pi^{-1}(\mathbf{b})$ simultaneously.

Theorem 5.3. *Let $\mathcal{F} \subset ker(\pi)$. The graphs $\pi^{-1}(\mathbf{b})_{\mathcal{F}}$ are connected for all $\mathbf{b} \in \mathbf{N}\mathcal{A}$ if and only if the set $\{\mathbf{x}^{\mathbf{v}^+} - \mathbf{x}^{\mathbf{v}^-} : \mathbf{v} \in \mathcal{F}\}$ generates the toric ideal $I_{\mathcal{A}}$.*

Proof: Let $\langle\mathcal{F}\rangle$ denote the ideal generated by $\{\mathbf{x}^{\mathbf{v}^+} - \mathbf{x}^{\mathbf{v}^-} : \mathbf{v} \in \mathcal{F}\}$. By Corollary 4.3, we have $\langle\mathcal{F}\rangle \subseteq I_{\mathcal{A}}$. We must show that equality holds if and only if all the graphs $\pi^{-1}(\mathbf{b})_{\mathcal{F}}$ are connected. We begin with the "if"-direction. Given any $\mathbf{u} \in ker(\pi)$, we must show that $\mathbf{x}^{\mathbf{u}^+} - \mathbf{x}^{\mathbf{u}^-}$ lies in $\langle\mathcal{F}\rangle$. Let $\mathbf{b} := \pi(\mathbf{u}^+) = \pi(\mathbf{u}^-)$. Since $\pi^{-1}(\mathbf{b})_{\mathcal{F}}$ is a connected graph, there exists a connecting path in that fiber: $\mathbf{u}^+ = \mathbf{u}^{(0)}, \mathbf{u}^{(1)}, \mathbf{u}^{(2)}, \ldots, \mathbf{u}^{(r-1)}, \mathbf{u}^{(r)} = \mathbf{u}^-$. This means that each binomial $\mathbf{x}^{\mathbf{u}^{(i-1)}} - \mathbf{x}^{\mathbf{u}^{(i)}}$ lies in $\langle\mathcal{F}\rangle$. Hence so does their sum

$$\mathbf{x}^{\mathbf{u}^+} - \mathbf{x}^{\mathbf{u}^-} \quad = \quad \sum_{i=1}^{r} (\mathbf{x}^{\mathbf{u}^{(i-1)}} - \mathbf{x}^{\mathbf{u}^{(i)}}) \quad \in \quad \langle\mathcal{F}\rangle.$$

For the "only-if" direction we assume that $\langle\,\mathcal{F}\,\rangle = I_{\mathcal{A}}$. Let $\mathbf{u}$ and $\mathbf{u}'$ be any two lattice points in the same fiber $\pi^{-1}(\mathbf{b})$. There exists a representation

$$\mathbf{x}^{\mathbf{u}} - \mathbf{x}^{\mathbf{u}'} \quad = \quad \sum_{i=1}^{N} \mathbf{x}^{\mathbf{w}_i} \cdot (\mathbf{x}^{\mathbf{v}_i^+} - \mathbf{x}^{\mathbf{v}_i^-}), \tag{5.6}$$

where each $\mathbf{v}_i = \mathbf{v}_i^+ - \mathbf{v}_i^-$ is a vector in $\mathcal{F}$, possibly occurring more than once in (5.6). If $N = 1$ then (5.6) is equivalent to $\mathbf{u} - \mathbf{u}' \in \mathcal{F}$, and $\mathbf{u}$ and $\mathbf{u}'$ are connected by an edge in $\pi^{-1}(\mathbf{b})_{\mathcal{F}}$. For $N > 1$ we wish to show that $\mathbf{u}$ and $\mathbf{u}'$ are connected by a path. We shall proceed by induction on N. The monomial $\mathbf{x}^{\mathbf{u}}$ is equal to one of the terms $\mathbf{x}^{\mathbf{w}_i}\mathbf{x}^{\mathbf{v}_i^+}$ or $\mathbf{x}^{\mathbf{w}_i}\mathbf{x}^{\mathbf{v}_i^-}$ appearing in the expansion of the right hand side of (5.6). After relabeling the sum we may assume that $\mathbf{u} = \mathbf{w}_1 + \mathbf{v}_1^+$. This implies that $\mathbf{u}$ and $\mathbf{w}_1 + {\mathbf{v}_1}^-$ are connected by an edge in $\pi^{-1}(\mathbf{b})_{\mathcal{F}}$. By deleting the first summand in (5.6), we get an expression for $\mathbf{x}^{\mathbf{w}_1 + {\mathbf{v}_1}^-} - \mathbf{x}^{\mathbf{u}'}$ which has length $N - 1$. By induction, $\mathbf{w}_1 + {\mathbf{v}_1}^-$ and $\mathbf{u}'$ are connected by a path in $\pi^{-1}(\mathbf{b})_{\mathcal{F}}$, and hence $\mathbf{u}$ and $\mathbf{u}'$ are connected as well. ∎

Example 5.1. (continued)
The toric ideal $I_{\mathcal{A}}$ is minimally generated by the 2×2-minors of an $s \times t$-matrix of indeterminates. We shall prove this in Proposition 5.4 below. The projective toric variety $\mathcal{V}(I_{\mathcal{A}})$ is the Segre embedding of $P^{r-1} \times P^{s-1}$ into P^{rs-1}.

For $s = t = 3$ the connecting moves corresponding to the 2×2-minors are:

$$\mathcal{F} = \left\{ \begin{pmatrix} +1 & -1 & 0 \\ -1 & +1 & 0 \\ 0 & 0 & 0 \end{pmatrix}, \begin{pmatrix} +1 & 0 & -1 \\ -1 & 0 & +1 \\ 0 & 0 & 0 \end{pmatrix}, \begin{pmatrix} 0 & +1 & -1 \\ 0 & -1 & +1 \\ 0 & 0 & 0 \end{pmatrix}, \right.$$
$$\begin{pmatrix} +1 & -1 & 0 \\ 0 & 0 & 0 \\ -1 & +1 & 0 \end{pmatrix}, \begin{pmatrix} +1 & 0 & -1 \\ 0 & 0 & 0 \\ -1 & 0 & +1 \end{pmatrix}, \begin{pmatrix} 0 & +1 & -1 \\ 0 & 0 & 0 \\ 0 & -1 & +1 \end{pmatrix},$$
$$\left. \begin{pmatrix} 0 & 0 & 0 \\ +1 & -1 & 0 \\ -1 & +1 & 0 \end{pmatrix}, \begin{pmatrix} 0 & 0 & 0 \\ +1 & 0 & -1 \\ -1 & 0 & +1 \end{pmatrix}, \begin{pmatrix} 0 & 0 & 0 \\ 0 & +1 & -1 \\ 0 & -1 & +1 \end{pmatrix} \right\}.$$

We invite the reader to draw the graph $\pi^{-1}(1,1,2;1,1,2)_{\mathcal{F}}$ and to verify that it is connected. Its node set is given in equation (5.2). Theorem 5.2 (together with Proposition 5.4 below) implies that all graphs $\pi^{-1}(r_1, r_2, r_3; c_1, c_2, c_3)_{\mathcal{F}}$ are connected. Note that this property ceases to hold if any one of the nine moves is omitted from $\mathcal{F}$. ∎

Proposition 5.4. *The toric ideal $I_{\mathcal{A}}$ for $\mathcal{A} = \{\mathbf{e}_i \oplus \mathbf{e}'_j\}$ is the kernel of the map*

$$\hat{\pi} \; : \; k[x_{11}, \ldots, x_{st}] \;\rightarrow\; k[y_1, \ldots, y_s, z_1, \ldots, z_t] \, , \quad x_{ij} \;\mapsto\; y_i z_j .$$

With respect to a suitable term order "$\prec$" the reduced Gröbner basis of $I_{\mathcal{A}}$ equals

$$\mathcal{G}_{\prec} \quad = \quad \{\, \underline{x_{il}x_{jk}} - x_{ik}x_{jl} \; : \; 1 \leq i < j \leq s, \, 1 \leq k < l \leq t \,\}. \tag{5.7}$$

Proof: It is easy to see that $\hat{\pi}$ is the lifting of the map π in (5.1), and that the 2×2-minors lie in $I_{\mathcal{A}}$. To show that they are a reduced Gröbner basis, we choose the reverse lexicographic term order induced from the row-wise variable ordering $x_{11} \prec x_{12} \prec \cdots \prec x_{st}$. This term order is called the *diagonal term order.* The initial terms are underlined in (5.7). We proceed by induction on $r+s$, the assertion being obvious for $r = s = 2$. Let $r+s \geq 5$ and suppose that $\mathcal{G}_\prec$ is not the reduced Gröbner basis. Then there exists $\mathbf{u} \in ker(\pi)$ such that neither $\mathbf{x}^{\mathbf{u}^+}$ nor $\mathbf{x}^{\mathbf{u}^-}$ is divisible by any of the underlined monomials. Let $(\mathbf{r}, \mathbf{c}) := \pi(\mathbf{u}^+) = \pi(\mathbf{u}^-)$. By Induction, our assertion holds for $(s-1) \times t$-matrices and for $s \times (t-1)$-matrices. Therefore all coordinates of $(\mathbf{r}, \mathbf{c})$ are non-zero; in particular, we have $r_1 \geq 1$ and $c_1 \geq 1$. Since $\mathbf{x}^{\mathbf{u}^+}$ is not divisible by any product $x_{1l}x_{j1}$ with $j, l \geq 2$, it follows that $\mathbf{x}^{\mathbf{u}^+}$ is divisible by x_{11}. The same reasoning shows that also $\mathbf{x}^{\mathbf{u}^-}$ is divisible by x_{11}. This is a contradiction because $\mathbf{u}^+$ and $\mathbf{u}^-$ have disjoint support. ∎

We next turn to the problem of integer programming. Let $\prec$ be any term order on $\mathbf{N}^n$. We define a directed graph (short: *digraph*) $\pi^{-1}(\mathbf{b})_{\mathcal{F},\prec}$ as follows. The underlying undirected graph is $\pi^{-1}(\mathbf{b})_{\mathcal{F}}$. An edge $(\mathbf{u}, \mathbf{u}')$ is directed from $\mathbf{u}$ to $\mathbf{u}'$ if $\mathbf{u}' \prec \mathbf{u}$.

Theorem 5.5. *Let $\mathcal{G} \subset ker(\pi)$ and $\prec$ any term order on $\mathbf{N}^n$. The directed graph $\pi^{-1}(\mathbf{b})_{\mathcal{G},\prec}$ has a unique sink for every $\mathbf{b} \in \mathbf{N}\mathcal{A}$ if and only if the set of binomials $\{\, \mathbf{x}^{\mathbf{v}^+} - \mathbf{x}^{\mathbf{v}^-} : \mathbf{v} \in \mathcal{G} \}$ is a Gröbner basis for the toric ideal $I_\mathcal{A}$ with respect to $\prec$.*

Proof: The directed edges in $\pi^{-1}(\mathbf{b})_{\mathcal{G},\prec}$ are the possible monomial reductions with respect to $\mathcal{G}$. Suppose $\mathcal{G}$ is a Gröbner basis. All monomials lying in the same fiber $\pi^{-1}(\mathbf{b})$ have the same normal form $\mathbf{x}^{\mathbf{u}}$ modulo $\mathcal{G}$. Then $\mathbf{u}$ is the unique sink in the digraph $\pi^{-1}(\mathbf{b})_{\mathcal{G},\prec}$. For the converse suppose that $\mathcal{G}$ is not a Gröbner basis. Then there exists $\mathbf{u} \in ker(\pi) \setminus \{0\}$ such that neither $\mathbf{x}^{\mathbf{u}^+}$ nor $\mathbf{x}^{\mathbf{u}^-}$ can be further reduced modulo $\mathcal{G}$. Let $\mathbf{b} := \pi(\mathbf{u}^+) = \pi(\mathbf{u}^-)$. Our assumption implies that $\mathbf{u}^+$ and $\mathbf{u}^-$ are distinct sinks in the digraph $\pi^{-1}(\mathbf{b})_{\mathcal{G},\prec}$. ∎

Recall from Chapter 1 that each cost vector $\omega \in \mathbf{R}^n$ can be refined to a term order $\prec_\omega$. We allow ω to have negative coordinates because $I_\mathcal{A}$ is homogeneous (cf. Proposition 1.12) with respect to some positive grading (e.g., take $deg(x_i) =$ *the coordinate sum of* $\mathbf{a}_i$).

Algorithm 5.6. *(Integer programming for a fixed matrix and cost function)*
`Input:` A $d \times n$-matrix $\mathcal{A}$ as above and a cost function $\omega \in \mathbf{R}^n$.
`Output:` An optimal point $\mathbf{u} \in \pi^{-1}(\mathbf{b})$ with $\mathbf{u} \cdot \omega$ minimal, for any given $\mathbf{b} \in \mathbf{N}\mathcal{A}$.

1. Compute the reduced Gröbner basis $\mathcal{G}_{\prec_\omega}$ for $I_\mathcal{A}$ with respect to $\prec_\omega$.
2. For any given right hand side vector $\mathbf{b} \in \mathbf{N}\mathcal{A}$ do:
   - 2.1 Find any feasible solution $\mathbf{v} \in \pi^{-1}(\mathbf{b})$.
   - 2.2 Compute the normal form $\mathbf{x}^{\mathbf{u}}$ of $\mathbf{x}^{\mathbf{v}}$ with respect to $\mathcal{G}_{\prec_\omega}$. Output $\mathbf{u}$.

Step 2.1 is analogous to PHASE-I in the simplex algorithm for linear programming. It can be done by computing any Gröbner basis as in Algorithm 4.5, and then reducing $\mathbf{t}^{\mathbf{b}}$ with respect to that Gröbner basis. It is often the case that a feasible solution $\mathbf{v}$ is given as part of the input, in which case step 2.1 can be omitted.

Example 5.1. (continued)
Let $s = t = 4$. The cost matrix ω in (5.5) defines a term order $\prec_\omega$ which is equivalent to the diagonal term order in Proposition 5.4. Hence $\mathcal{G}_{\prec_\omega}$ is just the set of 2×2-minors. We consider the table in (5.4) as a given feasible solution. It is

coded as a monomial:

$$\mathbf{x}^{\mathbf{v}} \;=\; x_{11}^{68}x_{12}^{119}x_{13}^{26}x_{14}^{7}\,x_{21}^{20}x_{22}^{84}x_{23}^{17}x_{24}^{94}\,x_{31}^{15}x_{32}^{54}x_{33}^{14}x_{34}^{10}\,x_{41}^{5}x_{42}^{29}x_{43}^{14}x_{44}^{16}.$$

The normal form of $\mathbf{x}^{\mathbf{v}}$ with respect to the Gröbner basis in Proposition 5.4 equals

$$\mathbf{x}^{\mathbf{u}} \;=\; x_{11}^{108}x_{12}^{112}x_{22}^{174}x_{23}^{41}x_{33}^{30}x_{34}^{63}x_{44}^{64}.$$

This is exactly how the optimal solution in (5.5) was found. ■

The enumeration problem for fibers $\pi^{-1}(\mathbf{b})$ of moderate size can be solved as follows.

Algorithm 5.7. *(Enumeration of fibers using Gröbner bases)*
`Input:` A $d \times n$-matrix $\mathcal{A}$ as above and a vector $\mathbf{b}$ in the semigroup $\mathbf{N}\mathcal{A}$.
`Output:` A list of all elements in the fiber $\pi^{-1}(\mathbf{b})$.

1. Compute any Gröbner basis $\mathcal{G}$ for $I_{\mathcal{A}}$.
2. Find any feasible solution $\mathbf{u}' \in \pi^{-1}(\mathbf{b})$.
3. By reducing $\mathbf{x}^{\mathbf{u}'}$ modulo $\mathcal{G}$, find the unique sink $\mathbf{u}''$ in the digraph $\pi^{-1}(\mathbf{b})_{\mathcal{G},\prec}$.
4. Now run a backward search of the digraph $\pi^{-1}(\mathbf{b})_{\mathcal{G},\prec}$, for instance as follows: First initialize `Active` $:= \{\mathbf{u}''\}$. `Passive` $:=$ the empty set.
5. While `Active` is non-empty do
   5.1. Choose any element $\mathbf{u} \in$ `Active`.
   5.2. For all $\mathbf{v} = \mathbf{v}^+ - \mathbf{v}^- \in \mathcal{G}$ (with $\mathbf{v}^+ \succ \mathbf{v}^-$) do
      5.2.1 if $\mathbf{u} - \mathbf{v}^- \geq 0$ and $\mathbf{u} + \mathbf{v} \notin$ `Passive` then `Active` $:=$ `Active` $\cup$ $\{\mathbf{u} + \mathbf{v}\}$.
   5.3. `Active` $:=$ `Active` $\setminus \{\mathbf{u}\}$. `Passive` $:=$ `Passive` $\cup \{\mathbf{u}\}$.
6. Output `Passive`.

One drawback of Algorithm 5.7 as presented is that the set `Active` can grow very large during the computation. This problem can be resolved by applying the "reverse search" technique of (Avis & Fukuda 1992). The reverse search variant of Algorithm 5.7 requires no intermediate storage whatsoever, and it runs in linear time in the size of the output.

Exercises:

(1) Compute the universal Gröbner basis for the ideal of 2×2-minors in Example 5.1.

(2) Compute the degree of the Segre embedding of $P^{r-1} \times P^{s-1}$ using Theorem 4.16. Here $Q = conv(\mathcal{A})$ is the product of a regular $(r-1)$-simplex with a regular $(s-1)$-simplex.

(3) The convex hull of the seven matrices in (5.2) is a four-dimensional polytope. Compute all faces of this polytope. How about the convex hull of $\pi^{-1}(42, 42, 5; 28, 30, 31)$?

(4) How many ways are there of expressing 10 dollars in terms of pennies, nickels, dimes and quarters, using exactly 100 coins ? Use Algorithm 5.7 to list all possibilities.

(5) Run Algorithm 5.2 on the data set (5.4). Among all 4×4-matrices with the same row and column sums, what is the average entry in the upper left hand corner ?

(6) Often the solution to an integer programming problem is not unique, and one might like to know all optimal solutions or all extreme optimal solutions.
   6.1 Give an algorithm for computing all optimal solutions to an integer programming problem. (Hint: Combine Algorithms 5.6 and 5.7).
   6.2 Give an algorithm for computing the vertices of $face_\omega(conv(\pi^{-1}(\mathbf{b})))$.

(7) Consider the following map between two polynomial rings each having 27 variables:

$$\Phi : k[x_{rst}] \to k[u_{rs}, v_{rt}, w_{st}], \quad x_{rst} \mapsto u_{rs} \cdot v_{rt} \cdot w_{st}. \qquad (1 \leq r, s, t \leq 3).$$

   7.1. The kernel of Φ is a toric ideal $I_\mathcal{A}$. Determine the configuration $\mathcal{A} \subset \mathbf{Z}^{27}$.
   7.2. Explain the integer program and sampling problem associated with $\mathcal{A}$.
   7.3. Find a minimal generating set and a Gröbner basis for $I_\mathcal{A}$.
   7.4. Compute the facets and the normalized volume of the polytope $Q = conv(\mathcal{A})$.

Notes:
The method of using toric ideals for sampling from conditional distributions was suggested in (Diaconis & Sturmfels 1993). Theorem 5.3 appears in that paper. Ideal generation criteria in terms of graphs equivalent to Theorem 5.3 can also be found in (Bresinsky 1988) and (Ollivier 1991). The application to integer programming has its origin in (Conti & Traverso 1991). It was further developed in (Pottier 1994), (Thomas 1994), (Sturmfels & Thomas 1994), (Sturmfels, Weismantel & Ziegler 1995), and (Urbaniak, Weismantel & Ziegler 1994). For the relations between Gröbner bases and existing methods in integer programming the reader may wish to consult these four articles. A combination of Algorithms 5.6 and 5.7 is applied in (Tayur, Thomas & Natraj 1995) to solve a class of stochastic integer programs arising in manufacturing. The question of describing the cardinality of $\pi^{-1}(\mathbf{b})$ as a function of $\mathbf{b}$ is addressed in (Sturmfels 1996).

CHAPTER 6

Primitive Partition Identities

In what follows we investigate generalizations of the well-known identity $1+1=2$. Fix a positive integer n. A *partition identity* is any identity of the form

$$a_1 + a_2 + a_3 + \cdots + a_k \quad = \quad b_1 + b_2 + b_3 + \cdots + b_l, \tag{6.1}$$

where $0 < a_i, b_j \leq n$ and all parts integers (generally not distinct). The number $k + l$ is called its *degree*. We call (6.1) *primitive* if there is no proper subidentity

$$a_{i_1} + a_{i_2} + \cdots + a_{i_r} \quad = \quad b_{j_1} + b_{j_2} + \cdots + b_{j_s}, \tag{6.2}$$

where $1 \leq r + s \leq k + l - 1$. Thus $1+3+3+3=5+5$ is a primitive partition identity of degree six and largest part equal to five.

To further illustrate our definition we list all primitive partition identities (ppi's) for $n = 5$. The ppi's for all smaller values of n appear in the beginning.

$1+1=2,\quad 2+2=1+3,\ 2+2+2=3+3,\ 1+2=3,\ 1+1+1=3,$
$3+3=2+4,\ 3+3+3=1+4+4,\ 3+3+3+3=4+4+4,\ 2+3=1+4,$
$2+3+3=4+4,\ 2+2=4,\ 1+3=4,\ 3+3=1+1+4,\ 1+1+2=4,$
$1+1+1+1=4,\quad 4+4=3+5,\ 4+4+4=2+5+5,$
$4+4+4+4=1+5+5+5,\quad 4+4+4+4+4=5+5+5+5,$
$3+4=2+5,\ 3+4+4=1+5+5,\ 3+4+4+4=5+5+5,\ 3+3=1+5,$
$3+3+4=5+5,\ 3+3+3=4+5,\ 3+3+3=2+2+5,\ 3+3+3+3=2+5+5,$
$3+3+3+3+3=5+5+5,\ 2+4=1+5,\ 2+4+4=5+5,\ 2+3=5,$
$2+2+4=3+5,\ 2+2+2=1+5,\ 2+2+2+4=5+5,\ 2+2+2+2=3+5,$
$2+2+2+2+2=5+5,\ 1+4=5,\ 1+3+3=2+5,\ 1+3+3+3=5+5,$
$4+4=1+2+5,\ 1+2+2=5,\ 3+4=1+1+5,\ 4+4+4=1+1+5+5,$
$1+1+3=5,\ 4+4=1+1+1+5,\ 1+1+1+2=5,\ 1+1+1+1+1=5.$

Table 6-1. *Primitive partition identities for $n \leq 5$.*

Using the technique to be described in the next chapter (Example 7.4), we extended this table up to $n = 13$. The results of these computations are summarized by cardinality:

n	2	3	4	5	6	7	8	9	10	11	12	13
# of ppi's	1	5	15	47	102	276	578	1261	2465	5362	9285	18900

Our first theorem concerns the maximum degree a ppi can have.

Theorem 6.1. *Any primitive partition identity (6.1) with largest part at most n satisfies $k + l \leq 2n - 1$. This degree bound is sharp. The ppi*

$$\underbrace{n+n+n+\cdots+n}_{n-1 \text{ terms}} \quad = \quad \underbrace{(n-1)+(n-1)+\cdots+(n-1)}_{n \text{ terms}}. \qquad (6.3)$$

is the unique primitive partition identity with $k + l = 2n - 1$.

Proof: Suppose that (6.1) is primitive. We may assume that n does not appear on the right hand side of (6.1). But it can appear on the left hand side. We run the following algorithm, starting with $x := 0$ and the multisets $\mathcal{P} := \{a_1, \ldots, a_k\}$ and $\mathcal{N} := \{b_1, \ldots, b_l\}$:

```
While 𝒫 ∪ 𝒩 is non-empty do
  if x ≥ 0
    then select an element ν ∈ 𝒩, set x := x − ν and 𝒩 := 𝒩 \ {ν}
    else select an element π ∈ 𝒫, set x := x + π and 𝒫 := 𝒫 \ {π}.
```

At each step in the while-loop the value of x is an integer between $1 - n$ and $n - 1$. Thus the total number of possible values for x is $2n - 1$. Since (6.1) is primitive, no value can be attained more than once. Otherwise a proper subidentity (6.2) is created whenever a value is reached for the second time. Therefore the total number of iterations in our loop is at most $2n - 1$, which proves the first part of Theorem 6.1.

The maximum degree $2n - 1$ can be attained only if all possible values for x are attained in the above loop. We add the requirement that in each step the largest element ν in $\mathcal{N}$ or π in $\mathcal{P}$ is to be selected. Then $\nu = n - 1$ in the first step. Otherwise the value $x = 1 - n$ will never be reached. The next time we enter the "then"-case, we must jump from $x = +1$ with $\nu = n - 1$. Otherwise the value $x = 2 - n$ will never be reached. The next time we enter the "then"-case, we must jump from $x = +2$ with $\nu = n - 1$. Otherwise the value $x = 3 - n$ will never be reached. Iterating this argument, we see that $b_1 = b_2 = \cdots = b_l = n - 1$ and $l = n$. This proves that (6.3) is the only primitive identity of maximum degree. ■

The upper bound in Theorem 6.1 can be strengthened as follows:

Corollary 6.2. *If (6.1) is a primitive partition identity, then*

$$k + l \;\leq\; max\{\, a_i \,:\, i = 1, \ldots, k\} \;+\; max\{\, b_j \,:\, j = 1, \ldots, l\}.$$

Proof: Let a_{i_0} be the maximum of the a_i's and let b_{j_0} be the maximum of the b_j's. In our algorithm in the proof of Theorem 6.1 the value of x is always an integer between $-b_{j_0}$ and $a_{i_0} - 1$. So, the number of possible values for x equals $a_{i_0} + b_{j_0}$, which is the right hand side of the claimed inequality. ■

What does all of this have to do with toric ideals ? The answer is simple:

Observation 6.3. *Let $d = 1$ and $\mathcal{A} = \{1, 2, 3, \ldots, n\}$. Then $x_{a_1} x_{a_2} \cdots x_{a_k} - x_{b_1} x_{b_2} \cdots x_{b_l}$ is a primitive binomial in $I_\mathcal{A}$ if and only if (6.1) is a primitive partition identity.*

Thus the ppi's are precisely the elements in the Graver basis $Gr_\mathcal{A}$ for the ideal $I_\mathcal{A} = kernel\big(k[x_1, \ldots, x_n] \to k[t],\, x_i \mapsto t^i\,\big)$. The ppi's in Table 6-1 are in binomial notation

$$x_1^2 - x_2\,,\; x_2^2 - x_1 x_3\,,\; x_2^3 - x_3^2,\; x_1 x_2 - x_3\,,\; x_1^3 - x_3\,,\; x_3^2 - x_2 x_4\,,\; x_3^3 - x_1 x_4^2,\; \ldots$$

The toric variety $\mathcal{V}(I_\mathcal{A})$ is the *affine rational normal curve* in k^n. In algebraic geometry it is more common to consider this curve in projective space. For the projective version we set $d = 2$ and $\mathcal{A} := \{(1,1),(1,2),\ldots,(1,n)\}$. The ideal $I_\mathcal{A}$ is homogeneous. The projective toric variety $\mathcal{V}(I_\mathcal{A})$ in P^{n-1} is the *rational normal curve of degree* $n-1$.

We say that the partition identity (6.1) is *homogeneous* if $k = l$. It is *homogeneous primitive* if, in addition, no proper subidentity (6.2) with $r = s$ exists. The homogeneous primitive partition identities (6.1) correspond to the Graver basis elements $x_{a_1}\cdots x_{a_k} - x_{b_1}\cdots x_{b_k}$ for $\mathcal{A} = \{(1,1),\ldots,(1,n)\}$. We list all homogeneous primitive partition identities for $n \le 6$:

$$
\begin{array}{l}
2+2=1+3,\\
3+3=2+4,\ 3+3+3=1+4+4,\ 2+3=1+4,\ 2+2+2=1+1+4,\\
4+4=3+5,\ 4+4+4=2+5+5,\ 4+4+4+4=1+5+5+5,\\
3+4=2+5,\ 3+4+4=1+5+5,\ 2+2+2+2=1+1+1+5,\\
2+4=1+5,\ 2+2+3=1+1+5,\ 1+4+4=2+2+5,\ 3+3+3=2+2+5,\\
3+3=1+5,\ 5+5=4+6,\ 5+5+5=3+6+6,\ 5+5+5+5=2+6+6+6,\\
\underline{5+5+5+5+5=1+6+6+6+6},\ 4+5=3+6,\ 4+5+5=2+6+6,\\
4+5+5+5=1+6+6+6,\ 4+4=2+6,\ 4+4+4+4=1+3+6+6,\\
4+4+4=3+3+6,\ 4+4+4=1+5+6,\ 3+5=2+6,\ 3+5+5=1+6+6,\\
3+3+4=2+2+6,\ 2+5+5=3+3+6,\ \underline{4+4+4+4+4=1+1+6+6+6}\\
2+5=1+6,\ 3+3+5=1+4+6,\ 3+4=1+6,\ 3+3+3+5=1+1+6+6,\\
3+3+3=1+2+6,\ 3+3+3+3=2+2+2+6,\ 3+3+3+3=1+1+4+6,\\
\underline{3+3+3+3+3=1+1+1+6+6},\ 2+4+4=1+3+6,\ 2+3+3=1+1+6,\\
2+4+4+4=1+1+6+6,\ 2+2+4=1+1+6,\ 2+2+2+3=1+1+1+6,\\
\underline{2+2+2+2+2=1+1+1+1+6},\ 1+5+5=2+3+6,\ 1+4+5=2+2+6,\\
1+5+5+5=2+2+6+6,\ 4+4+5=1+6+6,\ 1+1+5+5=2+2+2+6.
\end{array}
$$

Table 6-2. *Homogeneous primitive partition identities for $n \le 6$.*

Note that homogeneous primitive partition identities need not be primitive in the inhomogeneous sense. The identity $1+4+4 = 2+2+5$ shows this. Underlined in Table 6-2 are the four identities of maximum degree $10 = 2\cdot 6 - 2$.

Theorem 6.4. *Any primitive homogeneous partition identity (6.1) satisfies $k = l \le n-1$. This is sharp since*

$$\underbrace{1+1+\cdots+1}_{n-2\text{ terms}} + n \quad = \quad \underbrace{2+2+\cdots+2}_{n-1\text{ terms}}.$$

There are exactly $\phi(n-1)$ (the Euler phi-function) such maximal identities. For $n \ge 5$, there are $n + 2\phi(n-1) + 2\phi(n-2) - 6$ primitive identities with $k = l = n-2$.

Proof: We sort the left and right hand sides of (6.1) as follows:

$$a_1 \le a_2 \le a_3 \le \cdots \le a_k \qquad \text{and} \qquad b_1 \le b_2 \le b_3 \le \cdots \le b_k.$$

Consider the differences $\delta_i := a_i - b_i$, $i = 1, \ldots, k$. In the equation

$$\delta_1 + \delta_2 + \cdots + \delta_k \quad = \quad 0 \tag{6.4}$$

we separate the positive terms and the negative terms. The result is an inhomogeneous primitive partition identity of degree k. Let $\Delta_+ = max\{\delta_i : \delta_i > 0\}$ and $\Delta_- = max\{-\delta_j : \delta_j < 0\}$. By Corollary 6.2 applied to (6.4) we have $k \leq \Delta_+ + \Delta_-$.

We now choose indices i_0 and j_0 such that $b_{i_0} - a_{i_0} = \Delta_-$ and $a_{j_0} - b_{j_0} = \Delta_+$. We distinguish two cases. If $i_0 < j_0$ then

$$1 + \Delta_- \;\leq\; a_{i_0} + \Delta_- \;=\; b_{i_0} \;\leq\; b_{j_0} \;=\; a_{j_0} - \Delta_+ \;\leq\; n - \Delta_+. \tag{6.5}$$

If $i_0 > j_0$ then

$$n - \Delta_- \;\geq\; b_{i_0} - \Delta_- \;=\; a_{i_0} \;\geq\; a_{j_0} \;=\; b_{j_0} + \Delta_+ \;\geq\; 1 + \Delta_+. \tag{6.6}$$

In either case we have $\Delta_+ + \Delta_- \leq n - 1$, and therefore

$$\text{degree of (6.1)} \quad = \quad 2k \quad \leq \quad 2(\Delta_+ + \Delta_-) \quad \leq \quad 2n - 2. \tag{6.7}$$

This proves the first part of the claim.

To establish the second part of Theorem 6.4, we must characterize all primitive identities of maximal degree $2n - 2$. Let $e_1, e_2, \ldots$ denote the positive δ_i's and let $f_1, f_2, \ldots$ denote the negated negative δ_i's. Thus (6.4) is written as $e_1 + e_2 + \cdots = f_1 + f_2 + \cdots$. This is a primitive partition identity. We apply the add-subtract algorithm from the proof of Theorem 6.1. Since equality holds in (6.7), the variable x must attain each integer value between $-\Delta_-$ and $\Delta_+ - 1$ exactly once. In fact, this must be the case for <u>every</u> permutation of $e_1, e_2, \ldots$ and of $f_1, f_2, \ldots$ respectively.

We claim that $e_1 = e_2 = \cdots$ and $f_1 = f_2 = \cdots$. We assume the contrary, say $e_1 \neq e_2$. For our add-subtract algorithm we permute the e_i's so that e_2 is last and e_1 is second to last. Between the addition step with $\pi = e_1$ and the addition step with $\pi = e_2$ there may be several intermediate subtraction steps, say $\nu = f_1, f_2, \ldots, f_t$. Let $S \geq 0$ be the x-value immediately after the addition of $\pi = e_2$. At this point the variable x has visited each integer between $-\Delta_-$ and 0 and each integer between S and $\Delta_+ - 1$ exactly once, and it only has to run down from S to 0. The last negative value visited in this run equals $x = S - e_2$. We now change the positions of e_1 and e_2 in the permutation of the e_i's. Otherwise we leave the permutations untouched. Running the algorithm again, after the addition step with $\pi = e_2$ there is only one more negative value left to be visited. It is the same one as before, namely, $x = S - e_2$. Therefore we have precisely the same subtraction steps $\nu = f_1, f_2, \ldots, f_t$ between the addition of $\pi = e_2$ and the later addition of $\pi = e_1$. This implies $e_1 = e_2$ and the claim is proved.

The equations $e_1 = e_2 = \cdots$ and $f_1 = f_2 = \cdots$ imply that (6.1) has the form

$$\alpha + \cdots + \alpha + \beta + \cdots + \beta \quad = \quad \gamma + \cdots + \gamma + \delta + \cdots + \delta,$$

where the number of α's equals the number of γ's, the number of β's equals the number of δ's, and $1 \leq \alpha < \gamma \leq \delta < \beta \leq n - 1$. The assumption that equality holds in (6.7) translates into the equation $e_i + f_j = \gamma - \alpha + \beta - \delta = n$. This equation together with the previous chain of inequalities implies $\alpha = 1$, $\gamma = \delta$, and $\beta = n - 1$.

In summary, we conclude that every homogeneous primitive identity of maximum degree must have the form

$$\underbrace{1 + 1 + \cdots + 1}_{n-\ell-1 \text{ terms}} + \underbrace{n + n + \cdots + n}_{\ell \text{ terms}} \quad = \quad \underbrace{(\ell+1) + (\ell+1) + \cdots + (\ell+1)}_{n-1 \text{ terms}} \qquad (6.8)$$

for some integer ℓ between 1 and $n-1$. The homogeneous identity (6.8) is seen to be primitive if and only if $gcd(n-1-\ell, \ell) = 1 = gcd(n-1, \ell)$. The number of integers ℓ with these properties equals $\phi(n-1)$, the value of the Euler phi-function. A similar (but more complicated) argument applies to give the result we state for degree $2n-4$. ■

In the next table we present a count by degree of all homogeneous primitive partition identities for $n \leq 13$:

degree	4	6	8	10	12	14	16	18	20	22	24	*total #*
$n = 3$	1											1
$n = 4$	3	2										5
$n = 5$	7	7	2									16
$n = 6$	13	22	12	4								51
$n = 7$	22	54	36	13	2							127
$n = 8$	34	118	110	54	18	6						340
$n = 9$	50	230	276	155	60	23	4					798
$n = 10$	70	418	646	406	182	78	24	6				1830
$n = 11$	95	710	1374	965	462	207	74	25	4			3916
$n = 12$	125	1150	2788	2260	1228	602	264	108	34	10		8569
$n = 13$	161	1783	5286	4696	2656	1343	628	278	98	35	4	16968

Table 6-3. *Degree distribution of homogeneous primitive partition identities*

To extend this table for $n > 13$ is a nice benchmark problem for implementations of the Buchberger Algorithm for toric ideals. Our results have the following consequence for the degrees of certain Gröbner bases. By a *monomial curve* we mean the projective toric variety defined by $\mathcal{A} = \{(1, i_1), (1, i_2), \ldots, (1, i_r)\}$ for any integers $i_1, i_2, \ldots, i_r$.

Corollary 6.5. *The maximum degree in any reduced Gröbner basis of the ideal of a monomial curve X is bounded above by the degree of X.*

Proof: We may assume that $\mathcal{A} = \{(1, i_1), (1, i_2), \ldots, (1, i_r)\}$ where $1 = i_1 < i_2 < \cdots < i_r = n$ and the differences $i_2 - i_1, i_3 - i_2 \ldots, i_r - i_{r-1}$ are relatively prime integers. By Theorem 4.16, the degree of the curve $X = \mathcal{V}(I_\mathcal{A})$ equals $i_r - i_1 = n - 1$. Let $\mathbf{x}^{\mathbf{u}_+} - \mathbf{x}^{\mathbf{u}_-}$ be any element in the universal Gröbner basis for $\mathcal{A}$. Both $\mathbf{x}^{\mathbf{u}_+}$ and $\mathbf{x}^{\mathbf{u}_-}$ have the same degree k. By Proposition 4.13 (c), the binomial $\mathbf{x}^{\mathbf{u}_+} - \mathbf{x}^{\mathbf{u}_-}$ appears in the universal Gröbner basis for $\mathcal{A}' = \{(1,1), (1,2), \ldots, (1,n)\}$. By Lemma 4.6, it appears in the Graver basis for $\mathcal{A}'$. Theorem 6.4 now implies $k \leq n - 1 = deg(X)$. ■

The integer programming problem associated with $\mathcal{A} = \{1, 2, \ldots, n\}$ is the following *knapsack problem*. (For $\mathcal{A} = \{(1,1), \ldots, (1,n)\}$ there is an analogous version.)

$$\begin{aligned} &\text{Minimize} && \sum_{j=1}^{n} \omega_j \cdot x_j \\ &\text{subject to} && \sum_{j=1}^{n} j \cdot x_j = \beta, \quad x_i \text{ integral and } 0 \leq x_i \leq d_i \end{aligned} \tag{6.9}$$

where $\omega_1, \ldots, \omega_n, d_1, \ldots, d_n$ and β are parameters ranging over the positive integers. For details and other formulations of the knapsack problem see (Schrijver 1986, Section 16.6).

A feasible solution $(x_1, \ldots, x_n)$ to (6.9) can be written as a pair of partitions:

$$\begin{array}{c|c} \text{inside the knapsack} & \text{outside the knapsack} \\ \underbrace{1,1,\ldots,1}_{x_1} \; \underbrace{2,2,\ldots,2}_{x_2} \; \cdots \; \underbrace{n,n,\ldots,n}_{x_n} & \underbrace{1,1,\ldots,1}_{d_1-x_1} \; \underbrace{2,\ldots,2}_{d_2-x_2} \; \cdots \; \underbrace{n,\ldots,n}_{d_n-x_n} \end{array} \tag{6.10}$$

Each partition identity (6.1) gets directed by the cost functional $(\omega_1, \ldots, \omega_n)$ via

$$\begin{aligned} a_1, a_2, a_3, \ldots, a_k \;\; \rightarrow \;\; & b_1, b_2, b_3, \ldots, b_l \\ & \text{whenever } \omega_{a_1} + \cdots + \omega_{a_k} > \omega_{b_1} + \cdots + \omega_{b_l}, \end{aligned} \tag{6.11}$$

provided lexicographic tie breaking is used if a tie occurs. We say that (6.10) *can be improved along* (6.11) if $a_1, a_2, \ldots, a_k$ appear on the left side ("inside the knapsack") and $b_1, b_2, \ldots, b_k$ appear on the right side ("outside the knapsack"). In this case the feasible solution (6.10) can be improved by the exchange step (6.11). We claim that the primitive partition identities are a *universal test set* for the general knapsack problem (6.9).

Corollary 6.6. *Let the ω_i, d_i and β be arbitrary integers. A feasible solution (6.7) to (6.6) is <u>not</u> optimal if and only if it can be improved along some <u>primitive</u> partition identity.*

Proof: This follows essentially from Lemma 4.6, Algorithm 5.6 and Observation 6.3. However, some care must be taken because the integer program associated with the set $\mathcal{A} = \{1, 2, \ldots, n\}$ does not have the upper bound constraints $x_i \leq d_i$ in (6.9). To make the proof correct we take a sneak preview into Chapter 7. Consider the Lawrence lifting $\Lambda(\mathcal{A})$ defined in the first paragraph of Chapter 7. By adding non-negative slack variables y_i, so that $x_i + y_i = d_i$, we pass from the coefficient matrix $\mathcal{A}$ to the new matrix $\Lambda(\mathcal{A})$. Since the Graver bases of $\mathcal{A}$ and $\Lambda(\mathcal{A})$ coincide by equation (7.2), we see that the presence of the additional constraints $x_i \leq d_i$ does not harm the validity of Corollary 6.6. ∎

Exercises:

(1) Let $d = 1, n = 4, \mathcal{A} = \{3, 5, 8, 9\}$. Use Gröbner methods to answer the following:
   (a) Minimize $100x_1 + 10x_2 + x_3$ subject to $3x_1 + 5x_2 + 8x_3 + 9x_4 = 47$ and $x_i \in \mathbf{N}$.
   (b) List all elements in the fiber $\pi^{-1}(47)$.
   (c) The polytope $conv(\pi^{-1}(47))$ is 3-dimensional. Draw its edge graph.

(2) *(The Frobenius Problem)* Let $m_1, \ldots, m_r$ be relatively prime non-negative integers. There exists a largest integer N which cannot be written as a sum of the m_i. Give an algorithm (using Gröbner bases) for computing the number $N = N(m_1, \ldots, m_r)$.

(3) Describe the Gröbner fan of the toric ideal $I_\mathcal{A}$ for $\mathcal{A} = \{1, 2, \ldots, n\}$. Compute this fan explicitly for $n = 3$ and 4.

(4) Give lower and upper bounds for the number of primitive partition identities with largest part at most n. What is the asymptotic behavior of this function for $n \to \infty$?

(5) A primitive partition identity (6.1) is said to be *square-free* if the integers a_i and b_j are all distinct. Compute all square-free ppi's with largest part $n = 15$. Explain how your output can be used to solve a certain class of $0-1$-knapsack problems.

Notes:

The material in this chapter is drawn from (Diaconis, Graham & Sturmfels 1995). Gröbner basis methods for the knapsack problem are also studied in (Urbaniak, Weismantel & Ziegler 1994).

CHAPTER 7

Universal Gröbner Bases

In this chapter we present techniques for computing the Graver basis $Gr_{\mathcal{A}}$ and the universal Gröbner basis $\mathcal{U}_{\mathcal{A}}$ of the given matrix $\mathcal{A} \in \mathbf{Z}^{d\times n}$. To this end we consider the enlarged matrix $\Lambda(\mathcal{A}) = \begin{pmatrix} \mathcal{A} & \mathbf{0} \\ \mathbf{1} & \mathbf{1} \end{pmatrix}$ where $\mathbf{1}$ is the $n \times n$-identity matrix and $\mathbf{0}$ is the $d \times n$-zero matrix. The $(d+n) \times 2n$-matrix $\Lambda(\mathcal{A})$ is called the *Lawrence lifting* of $\mathcal{A}$. Any matrix of the form $\Lambda(\mathcal{A})$ is said to be of *Lawrence type.* This construction and terminology stems from the theory of oriented matroids (see Section 9.3 in (Björner et.al. 1993)).

The matrices $\mathcal{A}$ and $\Lambda(\mathcal{A})$ have isomorphic kernels: $ker(\Lambda(\mathcal{A})) = \{(\mathbf{u}, -\mathbf{u}) : \mathbf{u} \in ker(\mathcal{A})\}$. The toric ideal $I_{\Lambda(\mathcal{A})}$ is the homogeneous prime ideal

$$I_{\Lambda(\mathcal{A})} \quad = \quad \langle \mathbf{x}^{\mathbf{u}^+}\mathbf{y}^{\mathbf{u}^-} - \mathbf{x}^{\mathbf{u}^-}\mathbf{y}^{\mathbf{u}^+} \; : \; \mathbf{u} \in ker(\mathcal{A}) \rangle. \tag{7.1}$$

in the polynomial ring $k[\mathbf{x}, \mathbf{y}] = k[x_1, \ldots, x_n, y_1, \ldots, y_n]$.

Theorem 7.1. *For a Lawrence type matrix $\Lambda(\mathcal{A})$ the following sets of binomials coincide:*
(i) the Graver basis of $\Lambda(\mathcal{A})$,
(ii) the universal Gröbner basis of $\Lambda(\mathcal{A})$,
(iii) any reduced Gröbner basis of $I_{\Lambda(\mathcal{A})}$,
(iv) any minimal binomial generating set of $I_{\Lambda(\mathcal{A})}$ (up to scalar multiples).

Proof: A vector $\mathbf{u}$ in $ker(\mathcal{A})$ is primitive if and only if the corresponding vector $(\mathbf{u}, -\mathbf{u})$ in $ker(\Lambda(\mathcal{A}))$ is primitive. Therefore the Graver bases of $\mathcal{A}$ and $\Lambda(\mathcal{A})$ are related as follows:

$$Gr_{\Lambda(\mathcal{A})} \quad = \quad \{\mathbf{x}^{\mathbf{u}^+}\mathbf{y}^{\mathbf{u}^-} - \mathbf{x}^{\mathbf{u}^-}\mathbf{y}^{\mathbf{u}^+} \; : \; \mathbf{x}^{\mathbf{u}^+} - \mathbf{x}^{\mathbf{u}^-} \in Gr_{\mathcal{A}}\}. \tag{7.2}$$

Clearly, $Gr_{\Lambda(\mathcal{A})}$ is a generating set of $I_{\Lambda(\mathcal{A})}$, and it is a Gröbner basis with respect to every term order. We must show that $Gr_{\Lambda(\mathcal{A})}$ is the unique minimal generating set of $I_{\Lambda(\mathcal{A})}$. Here "unique" means up to replacing the generators by scalar multiples.

Choose any element $g := \mathbf{x}^{\mathbf{u}^+}\mathbf{y}^{\mathbf{u}^-} - \mathbf{x}^{\mathbf{u}^-}\mathbf{y}^{\mathbf{u}^+}$ in $Gr_{\Lambda(\mathcal{A})}$. Let B be the set of all binomials $\mathbf{x}^{\mathbf{v}^+}\mathbf{y}^{\mathbf{v}^-} - \mathbf{x}^{\mathbf{v}^-}\mathbf{y}^{\mathbf{v}^+}$ in $I_{\Lambda(\mathcal{A})}$ except g. Suppose that B generates $I_{\Lambda(\mathcal{A})}$. Then g can be written as a polynomial linear combination of elements in B. This implies that there is a binomial $\mathbf{x}^{\mathbf{v}^+}\mathbf{y}^{\mathbf{v}^-} - \mathbf{x}^{\mathbf{v}^-}\mathbf{y}^{\mathbf{v}^+}$ in B one of whose terms divides $\mathbf{x}^{\mathbf{u}^+}\mathbf{y}^{\mathbf{u}^-}$. After replacing $\mathbf{v}$ by $-\mathbf{v}$ if necessary, we may assume that $\mathbf{x}^{\mathbf{v}^+}\mathbf{y}^{\mathbf{v}^-}$ divides $\mathbf{x}^{\mathbf{u}^+}\mathbf{y}^{\mathbf{u}^-}$. But this means that $\mathbf{u}$ is not primitive in $ker(\mathcal{A})$, which is a contradiction. Therefore some non-zero scalar multiple of g must appear in any minimal binomial generating set of $I_{\Lambda(\mathcal{A})}$. ∎

Theorem 7.1 suggests the following algorithm for computing Graver bases:

Algorithm 7.2. *(How to compute the Graver basis $Gr_{\mathcal{A}}$ of an integer matrix $\mathcal{A}$)*

1. Choose any term order on $k[\mathbf{x}, \mathbf{y}]$. Compute the reduced Gröbner basis $\mathcal{G}$ of $I_{\Lambda(\mathcal{A})}$.
2. Substitute $y_1, \ldots, y_n \mapsto 1$ in $\mathcal{G}$. The resulting subset of $k[\mathbf{x}]$ is the Graver basis $Gr_{\mathcal{A}}$.

The correctness of this algorithm is a corollary of Theorem 7.1. Algorithm 7.2 is extremely useful for explicit computations. The main point is that, in order to compute the Graver basis of $\mathcal{A}$, one only needs to compute a single reduced Gröbner basis for its Lawrence lifting $\Lambda(\mathcal{A})$. Step 1 of Algorithm 7.2 can be executed by applying Algorithm 4.5 to the Lawrence matrix $\Lambda(\mathcal{A})$. If each vector in $\mathcal{A} = \{\mathbf{a}_1, \mathbf{a}_2, \ldots, \mathbf{a}_n\}$ is non-negative, then it suffices to eliminate the variables $\mathbf{t} = (t_1, \ldots, t_d)$ from the binomial ideal

$$I \quad = \quad \langle x_1 - y_1 \mathbf{t}^{\mathbf{a}_1},\, x_2 - y_2 \mathbf{t}^{\mathbf{a}_2},\, \cdots\, x_n - y_n \mathbf{t}^{\mathbf{a}_n} \rangle \quad \subset \quad k[\mathbf{t}, \mathbf{x}, \mathbf{y}]. \tag{7.3}$$

Any generating set of $I \cap k[\mathbf{x}, \mathbf{y}] = I_{\Lambda(\mathcal{A})}$ automatically contains the Graver basis.

Example 7.3. *(How to compute all primitive partition identities (6.1))*
Compute the minimal generators of the elimination ideal

$$\langle x_1 - y_1 t,\, x_2 - y_2 t^2,\, x_3 - y_3 t^3,\, \ldots\,,\, x_n - y_n t^n \rangle \;\cap\; k[x_1, x_2, \ldots, x_n, y_1, y_2, \ldots, y_n],$$

and write them in the form $x_{a_1} x_{a_2} \cdots x_{a_k} y_{b_1} y_{b_2} \cdots y_{b_l} - y_{a_1} y_{a_2} \cdots y_{a_k} x_{b_1} x_{b_2} \cdots x_{b_l}$.
■

The universal Gröbner basis $\mathcal{U}_{\mathcal{A}}$ is a (generally proper) subset of the Graver basis $Gr_{\mathcal{A}}$. We next present a procedure for computing $\mathcal{U}_{\mathcal{A}}$ from $Gr_{\mathcal{A}}$. For the rest of this chapter we shall assume that each $\mathbf{a}_i$ is non-zero and non-negative. This implies that the toric ideal $I_{\mathcal{A}}$ is positively graded and hence the Gröbner region of $I_{\mathcal{A}}$ is all of $\mathbf{R}^n$ (Proposition 1.12).

We fix a tie breaking term order $\prec$. For $\omega \in \mathbf{R}^n$ let $\mathcal{G}_\omega$ denote the reduced Gröbner basis of $I_{\mathcal{A}}$ with respect to $\prec_\omega$. By varying ω we get all possible reduced Gröbner bases of $I_{\mathcal{A}}$. This is guaranteed by Proposition 1.11. Hence in (7.4) below $\mathcal{G}_\omega$ runs over distinct reduced Gröbner bases of $I_{\mathcal{A}}$. For $\mathbf{u} \in ker(\pi)$ we define

$$\begin{aligned} C_+[\mathbf{u}] \;&:=\; \{\omega \in \mathbf{R}^n \;:\; \omega \cdot \mathbf{u} > 0 \text{ and } \mathbf{x}^{\mathbf{u}^+} - \mathbf{x}^{\mathbf{u}^-} \in \mathcal{G}_\omega \} \\ \text{and} \quad C_-[\mathbf{u}] \;&:=\; \{\omega \in \mathbf{R}^n \;:\; \omega \cdot \mathbf{u} < 0 \text{ and } \mathbf{x}^{\mathbf{u}^+} - \mathbf{x}^{\mathbf{u}^-} \in \mathcal{G}_\omega \}. \end{aligned} \tag{7.4}$$

The vector $\mathbf{u}$ (or its corresponding binomial) lies in the universal Gröbner basis $\mathcal{U}_{\mathcal{A}}$ if and only if $C_+[\mathbf{u}] \cup C_-[\mathbf{u}] \neq \emptyset$. For $\mathbf{v} \in \mathbf{N}^n$ we introduce the open convex polyhedral cone

$$\mathcal{M}(\mathbf{v}) \;:=\; \{\omega \in \mathbf{R}^n \;:\; \omega \cdot \mathbf{v} < \omega \cdot \mathbf{w} \text{ for all } \mathbf{w} \in \pi^{-1}(\pi(\mathbf{v})) \setminus \{\mathbf{v}\} \}. \tag{7.5}$$

Note that the set of inequalities in (7.5) is finite because of our hypothesis that $I_{\mathcal{A}}$ is positively graded. Using the notation for normal cones in Chapter 2, we have $\mathcal{M}(\mathbf{v}) = -int \mathcal{N}_P(F)$, where F denotes the largest face of $P := conv(\pi^{-1}(\pi(\mathbf{v}))$ containing $\mathbf{v}$. In particular, the cone $\mathcal{M}(\mathbf{v})$ is empty unless $\mathbf{v}$ is a vertex of its fiber. We note the following reformulation:

$$\mathcal{M}(\mathbf{v}) \;=\; \{\omega \in \mathbf{R}^n \;:\; \mathbf{x}^{\mathbf{v}} \text{ does not lie in the initial ideal } in_\omega(I_{\mathcal{A}}) \}. \tag{7.6}$$

Lemma 7.4. *The cone $C_+[\mathbf{u}]$ equals the intersection*

$$\mathcal{M}(\mathbf{u}^-) \cap \bigcap_{i \in supp(\mathbf{u}^+)} \mathcal{M}(\mathbf{u}^+ - \mathbf{e}_i).$$

Proof: A cost vector $w \in \mathbf{R}^n$ belongs to $C_+[\mathbf{u}]$ if and only if $\mathbf{x}^{\mathbf{u}^+} - x^{\mathbf{u}^-}$ appears with initial term $x^{\mathbf{u}^+}$ in the reduced Gröbner basis $\mathcal{G}_\omega$. This holds if and only if $\mathbf{x}^{\mathbf{u}^-}$ is standard, $\mathbf{x}^{\mathbf{u}^+}$ is non-standard, and every proper factor of $\mathbf{x}^{\mathbf{u}^+}$ is standard (with respect to $\prec_\omega$). This is equivalent to $\omega \in \mathcal{M}(\mathbf{u}^-)$ and $\omega \in \mathcal{M}(\mathbf{u}^+ - \mathbf{e}_i)$ for all $i \in supp(\mathbf{u}^+)$. ∎

Corollary 7.5. *The set of all cost vectors ω which have a fixed binomial $\mathbf{x}^{\mathbf{u}^+} - \mathbf{x}^{\mathbf{u}^-}$ with fixed initial term $\mathbf{x}^{\mathbf{u}^+}$ in their reduced Gröbner basis is an open convex cone in $\mathbf{R}^n$.*

The punch line is that the cones $\mathcal{M}(\mathbf{v})$ can be computed from the Graver basis:

$$\begin{aligned} \mathcal{M}(\mathbf{v}) \; = \; \{\omega \in \mathbf{R}^n : \; & \omega \cdot \mathbf{u} > 0 \text{ for all } \mathbf{x}^{\mathbf{u}^+} - \mathbf{x}^{\mathbf{u}^-} \in Gr_\mathcal{A} \\ & \text{such that } \mathbf{x}^{\mathbf{u}^-} \text{ divides } \mathbf{x}^{\mathbf{v}}\}. \end{aligned} \tag{7.7}$$

This follows from (7.6) and the Gröbner basis property:

$$in_\omega(I_\mathcal{A}) \; = \; \langle in_\omega(f) : f \in Gr_\mathcal{A} \rangle.$$

Algorithm 7.6 *(How to compute the universal Gröbner basis $\mathcal{U}_\mathcal{A}$).*

1. Compute the Graver basis $Gr_\mathcal{A}$ using Algorithm 7.2.
2. For each binomial $\mathbf{x}^{\mathbf{u}^+} - \mathbf{x}^{\mathbf{u}^-}$ in $Gr_\mathcal{A}$ do:
   2.1. Compute the cones $C_+[\mathbf{u}]$ and $C_-[\mathbf{u}]$ using Lemma 7.4 and formula (7.7).
   2.2. The binomial $\mathbf{x}^{\mathbf{u}^+} - \mathbf{x}^{\mathbf{u}^-}$ is in $\mathcal{U}_\mathcal{A}$ if and only if $C_+[\mathbf{u}] \cup C_-[\mathbf{u}]$ is non-empty.

Example 7.7. Let $d = 3, n = 6$, $\mathcal{A} = \{(2,0,0),(1,1,0),(1,0,1),(0,2,0),(0,1,1),(0,0,2)\}$. The ideal $I_\mathcal{A}$ defines the *Veronese surface* in P^5. Using Algorithm 7.2 we find its Graver basis

$$\begin{aligned} Gr_\mathcal{A} = \{ & x_1x_4 - x_2^2,\, x_1x_6 - x_3^2,\, x_4x_6 - x_5^2,\, x_1x_5 - x_2x_3,\, x_4x_3 - x_2x_5, \\ & x_6x_2 - x_3x_5,\, x_1x_5^2 - x_2^2x_6,\, x_4x_3^2 - x_5^2x_1,\, x_6x_2^2 - x_3^2x_4,\, x_1x_4x_6 - x_2x_3x_5\}. \end{aligned}$$

We shall prove that

$$Gr_\mathcal{A} \setminus \mathcal{U}_\mathcal{A} \quad = \quad \{x_1x_4x_6 - x_2x_3x_5\}. \tag{7.8}$$

Using Lemma 4.8 and formula (4.5), it can be seen that the first nine binomials in $Gr_\mathcal{A}$ are precisely the circuits of $\mathcal{A}$. (Incidentally, finding these nine relations was also the task of Exercise (3) in Chapter 1.) In view of the inclusion $\mathcal{C}_\mathcal{A} \subseteq \mathcal{U}_\mathcal{A}$ in Proposition 4.11, it suffices to show that the binomial $x_1x_4x_6 - x_2x_3x_5$ is not in $\mathcal{U}_\mathcal{A}$. Consider its exponent vector $\mathbf{u} := (1,-1,-1,1,-1,1)$. By formula (7.7), we have $\mathcal{M}(\mathbf{u}^-) = \{\omega \in \mathbf{R}^6 : \omega_1 + \omega_5 > \omega_2 + \omega_3,\, \omega_2 + \omega_6 > \omega_3 + \omega_5,\, \omega_3 + \omega_4 > \omega_2 + \omega_5\}$, $\mathcal{M}(\mathbf{u}^+ - \mathbf{e}_1) = \{\omega \in \mathbf{R}^6 : 2\omega_5 > \omega_4 + \omega_6\}$, $\mathcal{M}(\mathbf{u}^+ - \mathbf{e}_4) = \{\omega \in \mathbf{R}^6 : 2\omega_3 > \omega_1 + \omega_6\}$ and $\mathcal{M}(\mathbf{u}^+ - \mathbf{e}_6) = \{\omega \in \mathbf{R}^6 : 2\omega_2 > \omega_1 + \omega_4\}$. The intersection of these four cones is easily seen to be empty, so that $C_+[\mathbf{u}] = \emptyset$. Reversing the roles of $\mathbf{u}^+$ and $\mathbf{u}^-$ we similarly find that $C_-[\mathbf{u}] = \emptyset$. This proves the claim (7.8). ∎

Our next result is a geometric characterization of the universal Gröbner basis. We say that an integer vector $\mathbf{u} \in \mathbf{Z}^n$ is *relatively prime* if its coordinates are relatively prime.

Theorem 7.8. *A relatively prime vector* $\mathbf{u} \in ker(\pi)$ *lies in the universal Gröbner basis* $\mathcal{U}_A$ *if and only if the line segment* $[\mathbf{u}^+, \mathbf{u}^-]$ *is an edge of the polytope* $conv(\pi^{-1}(\pi(\mathbf{u}^+))$.

Proof (if): Suppose $\mathbf{u}$ is relatively prime and $[\mathbf{u}^+, \mathbf{u}^-]$ an edge of $conv(\pi^{-1}(\pi(\mathbf{u}^+))$. There exists $\omega \in \mathbf{R}^n$ with

$$\omega \cdot \mathbf{u}^- < \omega \cdot \mathbf{u}^+ < \omega \cdot \mathbf{v} \quad \text{for all} \quad \mathbf{v} \in \pi^{-1}(\pi(\mathbf{u}^+)) \setminus \{\mathbf{u}^+, \mathbf{u}^-\}. \tag{7.9}$$

This implies $\omega \in \mathcal{M}(\mathbf{u}^-)$. In view of Lemma 7.4, it suffices to show that ω lies in $\bigcap_{i \in supp(\mathbf{u}^+)} \mathcal{M}(\mathbf{u}^+ - \mathbf{e}_i)$. Suppose not, say, $\omega \notin \mathcal{M}(\mathbf{u}^+ - \mathbf{e}_i)$. Then there exists $\mathbf{v} \in \pi^{-1}(\pi(\mathbf{u}^+ - \mathbf{e}_i)) \backslash \{\mathbf{u}^+ - \mathbf{e}_i\}$ such that $\omega \cdot \mathbf{v} \leq \omega \cdot (\mathbf{u}^+ - \mathbf{e}_i)$. This implies $\omega \cdot (\mathbf{v} + \mathbf{e}_i) \leq \omega \cdot \mathbf{u}^+$, and using (7.9) we conclude that $\mathbf{v} + \mathbf{e}_i = \mathbf{u}^-$. This is a contradiction to the fact that $\mathbf{u}^+$ and $\mathbf{u}^-$ have disjoint support. ∎

The argument just presented implies the following fact.

Corollary 7.9. *For every binomial* $f = \mathbf{x}^{\mathbf{u}^+} - \mathbf{x}^{\mathbf{u}^-}$ *in* $\mathcal{U}_A$ *there exist two term orders such that* f *appears with different initial terms in the two reduced Gröbner bases of* I_A.

For the proof of the only-if direction of Theorem 7.8 we need the following lemma.

Lemma 7.10. *Let* $\prec$ *be any term order, let* $\mathbf{x}^{\mathbf{u}}$ *be a minimal generator of the initial ideal* $in_{\prec}(I_A)$, *and let* $\mathbf{v}$ *be an element in* $\pi^{-1}(\pi(\mathbf{u}))$ *such that* $\mathbf{u} \succ \mathbf{v}$. *Then* $supp(\mathbf{u}) \cap supp(\mathbf{v}) = \emptyset$.

Proof: Suppose $i \in supp(\mathbf{u}) \cap supp(\mathbf{v})$ and $\mathbf{u} \succ \mathbf{v}$ and $\pi(\mathbf{u}) = \pi(\mathbf{v})$. Then $\mathbf{u} - \mathbf{e}_i$ and $\mathbf{v} - \mathbf{e}_i$ are non-negative and in the same fiber, and $\mathbf{u} - \mathbf{e}_i \succ \mathbf{v} - \mathbf{e}_i$. This implies that $\mathbf{x}^{\mathbf{u}-\mathbf{e}_i} = \mathbf{x}^{\mathbf{u}}/x_i$ lies in the initial ideal $in_{\prec}(I_A)$. This is a contradiction to our assumption that $\mathbf{x}^{\mathbf{u}}$ is a minimal generator. ∎

Proof of Theorem 7.8 (only-if): Suppose that $\mathbf{x}^{\mathbf{u}^+} - \mathbf{x}^{\mathbf{u}^-}$ appears with initial term $\mathbf{x}^{\mathbf{u}^+}$ in the reduced Gröbner basis of I_A with respect to $\prec$. Clearly, $\mathbf{u}$ must be a relatively prime lattice vector.

We had assumed that I_A is positively graded. This implies that both $\mathbf{u}^+$ and $\mathbf{u}^-$ are non-zero vectors. We must show that $conv\{\mathbf{u}^+, \mathbf{u}^-\}$ is an edge of its fiber. Let $\omega \in \mathbf{R}^n$ be a non-negative weight vector representing $\prec$. After replacing ω by a nearby vector if necessary, we may assume that every coordinate of ω is positive and that the linear functional $\mathbf{v} \mapsto \omega \cdot \mathbf{v}$ separates the points in $\pi^{-1}(\pi(\mathbf{u}^+))$. These two hypotheses are crucial in what follows.

Let ω' be the restriction of ω to the complement of $supp(\mathbf{u}^+)$, that is, $\omega'_i = 0$ if $u_i > 0$ and $\omega'_i = \omega_i$ if $u_i \leq 0$. Our hypotheses imply

$$0 = \omega' \cdot \mathbf{u}^+ < \omega' \cdot \mathbf{u}^- = \omega \cdot \mathbf{u}^- < \omega \cdot \mathbf{u}^+.$$

We define the weight vector

$$\omega'' \quad := \quad (\omega \cdot \mathbf{u}) \cdot \omega' - (\omega' \cdot \mathbf{u}) \cdot \omega,$$

which has the property $\omega''\mathbf{u}^+ = \omega''\mathbf{u}^-$. Note that all coordinates of ω'' are positive, because $\omega \cdot \mathbf{u} > 0$ and $\omega' \cdot \mathbf{u} < 0$. In order to prove that $conv\{\mathbf{u}^+, \mathbf{u}^-\}$ is an edge, it suffices to show that $\omega''\mathbf{u}^+ < \omega''\mathbf{v}$ for all $\mathbf{v} \in \mathbf{N}^n \backslash \{\mathbf{u}^-, \mathbf{u}^+\}$ with $\pi(\mathbf{v}) = \pi(\mathbf{u}^+)$.

We distinguish two cases. First suppose that $\omega\mathbf{v} < \omega\mathbf{u}^+$. Then $supp(\mathbf{u}^+)$ and $supp(\mathbf{v})$ are disjoint by Lemma 7.10, and hence $\omega\mathbf{v} = \omega'\mathbf{v}$. This implies

$$\omega''\mathbf{v} \;=\; \big(\omega\mathbf{u} - \omega'\mathbf{u}\big) \cdot \omega\mathbf{v} \;>\; \big(\omega\mathbf{u} - \omega'\mathbf{u}\big) \cdot \omega\mathbf{u}^- \;=\; \omega''\mathbf{u}^- \;=\; \omega''\mathbf{u}^+.$$

Next consider the case $\omega\mathbf{v} > \omega\mathbf{u}^+$. Then we have

$$\omega''\mathbf{v} \;>\; (\omega\mathbf{u}) \cdot \omega'\mathbf{v} - (\omega'\mathbf{u}) \cdot \omega\mathbf{u}^+ \;>\; -(\omega'\mathbf{u}) \cdot \omega\mathbf{u}^+ \;=\; \omega''\mathbf{u}^+.$$

This completes the proof. ∎

Theorem 7.8 gives rise to an alternative algorithm for computing the universal Gröbner basis from the Graver basis $Gr_{\mathcal{A}}$. Namely, for each binomial in $Gr_{\mathcal{A}}$ check whether it is an edge of its fiber, which amounts to solving a linear programming problem. This approach is particularly useful as a tool for proving that a Graver basis element does not lie in $\mathcal{U}_{\mathcal{A}}$.

Example 7.7. (continued) The fiber of the binomial $x_1x_4x_6 - x_2x_3x_5 \in Gr_{\mathcal{A}}$ contains precisely five lattice points. They form the vertices of a 3-dimensional bipyramid (see Figure 7-1). The line segment $conv\{(1,0,0,1,0,1), (0,1,1,0,1,0)\}$ is the diagonal of this bipyramidal fiber. This proves that this binomial is not in the universal Gröbner basis $\mathcal{U}_{\mathcal{A}}$. ∎

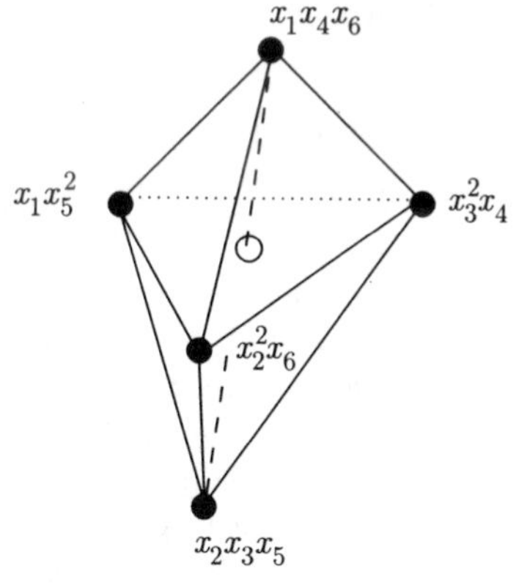

Figure 7-1. Bipyramidal fiber.

Example 7.11. *(A Graver degree which is not a Gröbner degree)*
Let $d = 1, n = 4$, and $\mathcal{A} = \{15, 20, 23, 24\}$. The *degree* of a monomial $x_1^{u_1}x_2^{u_2}x_3^{u_3}x_4^{u_4}$ is defined to be $15u_1+20u_2+23u_3+24u_4$. The ideal $I_{\mathcal{A}}$ is homogeneous with respect to this grading. An integer $m \geq 0$ is called a *Graver degree* (resp. *Gröbner degree*) if there exists an element in $Gr_{\mathcal{A}}$ (resp. in $\mathcal{U}_{\mathcal{A}}$) having degree m.

When computing the Graver basis using the method in Example 7.3, we find that $Gr_{\mathcal{A}}$ contains a unique binomial of degree 138. This binomial is $x_1^2x_2^3x_4^2 - x_3^6$. Consider the following convex combination of elements in $\pi^{-1}(138)$:

$$\frac{1}{4}(0,0,6,0) + \frac{3}{4}(2,3,0,2) \;\;=\;\; \frac{1}{4}(5,2,1,0) + \frac{1}{4}(1,5,1,0) + \frac{1}{2}(0,1,2,3).$$

This shows that $conv\{(2,3,0,2),(0,0,6,0)\}$, the segment corresponding to our binomial, is not an edge of its fiber $conv(\pi^{-1}(138))$. Using Theorem 7.8, we see that $x_1^2x_2^3x_4^2 - x_3^6$ does not lie in $\mathcal{U}_\mathcal{A}$. We conclude that 138 is a Graver degree but not a Gröbner degree. ∎

The concepts of Gröbner degrees and Graver degrees extend naturally to any $\mathcal{A} \subset \mathbf{N}^d$. A vector $\mathbf{b} \in \mathbf{N}\mathcal{A}$ is called a *Graver degree* (resp. *Gröbner degree*) if there exists a binomial $\mathbf{x}^{\mathbf{u}^+} - \mathbf{x}^{\mathbf{u}^-}$ in $Gr_\mathcal{A}$ (resp. in $\mathcal{U}_\mathcal{A}$) such that $\pi(\mathbf{u}^+) = \pi(\mathbf{u}^-) = \mathbf{b}$. If $\mathbf{b}$ is a Gröbner degree, then we call the polytope $conv(\pi^{-1}(\mathbf{b}))$ a *Gröbner fiber.*

There is a natural partial order on the semigroup $\mathbf{N}\mathcal{A}$. Namely, for two elements $\mathbf{b}, \mathbf{b}'$ in $\mathbf{N}\mathcal{A}$, we set $\mathbf{b}' \leq \mathbf{b}$ if and only if $\mathbf{b} - \mathbf{b}' \in \mathbf{N}\mathcal{A}$. This is well-defined since it is assumed that $\mathbf{N}\mathcal{A} \cap -\mathbf{N}\mathcal{A} = \{0\}$. In this partial order on all fibers the Gröbner fibers are characterized as follows.

Corollary 7.12. *A fiber $conv(\pi^{-1}(\mathbf{b}))$ is a Gröbner fiber if and only if it has an edge which is not parallel to any edge of a different fiber $conv(\pi^{-1}(\mathbf{b}'))$ with $\mathbf{b}' \leq \mathbf{b}$.*

Proof: We first prove the only-if direction. Let $conv(\pi^{-1}(\mathbf{b}))$ be a Gröbner fiber. By Theorem 7.8, it has a relatively prime edge of the form $[\mathbf{u}^+, \mathbf{u}^-]$, corresponding to an element of $\mathcal{U}_\mathcal{A}$. Suppose this segment were parallel to an edge $[\mathbf{v}^+, \mathbf{v}^-]$ of a different fiber $conv(\pi^{-1}(\mathbf{b}'))$ with $\mathbf{b}' \leq \mathbf{b}$. This implies that $\mathbf{v}^+ - \mathbf{v}^- = m \cdot (\mathbf{u}^+ - \mathbf{u}^-)$ for some non-zero integer m. After reversing the sign of $\mathbf{v}$, we may assume $m \geq 1$. We conclude that $\mathbf{v}^+ = m \cdot \mathbf{u}^+$ and therefore $\mathbf{b}' = \pi(\mathbf{v}^+) = m \cdot \pi(\mathbf{u}^+) = m \cdot \mathbf{b}$. This is a contradiction to the assumption $\mathbf{b}' \leq \mathbf{b}$. The proof of the if-direction is straightforward by reversing the argument, using the reverse implication in Theorem 7.8. ∎

This proof shows that the condition $\mathbf{b}' \leq \mathbf{b}$ in Corollary 7.12 can be weakened to $||\mathbf{b}'|| \leq ||\mathbf{b}||$ for any norm $|| \cdot ||$.

We next note that Gröbner fibers can have arbitrarily many vertices.

Example 7.13. *(Gröbner fibers with many vertices)*
This example is based on Remark 18.1 in (Schrijver 1986). Let ϕ_r denote the r-th Fibonacci number, which is defined recursively by $\phi_0 := 0$, $\phi_1 := 1$, $\phi_r := \phi_{r-2} + \phi_{r-1}$. Consider the 1×4-matrix $\mathcal{A}_r := [\, \phi_{2r}\,,\, \phi_{2r+1},\, 1\,,\, \phi_{2r+1}^2 - 1\,]$. Consider the fiber of $\mathcal{A}_r$ over $b_r = \phi_{2r+1}^2 - 1$. This is a Gröbner fiber because it is the fiber of the circuit $(\, 0,\, 0,\, 1 - \phi_{2r+1}^2,\, 1\,)$. The set of points with last coordinate zero is a facet of this fiber. It is a polygon isomorphic to the convex hull of all non-negative lattice points (x, y) with $\phi_{2r} \cdot x + \phi_{2r+1} \cdot y \leq \phi_{2r+1}^2 - 1$. This lattice polygon has $r+3$ vertices. We conclude that the b_r-fiber of $\mathcal{A}_r$ is a Gröbner fiber with at least $r+4$ vertices. ∎

The encoding of the lattice polygon as a facet of a 3-polytope in Example 7.13 is a special case of the following general construction.

Proposition 7.14. *Every lattice polytope appears as a facet of some Gröbner fiber.*

Proof: Every $(n-d)$-dimensional lattice polytope can be written as a fiber $conv(\pi^{-1}(\mathbf{b}))$ for some matrix $\mathcal{A} \in \mathbf{N}^{d \times n}$ of maximal row rank and some $\mathbf{b} \in \mathbf{N}^d$. This polytope is isomorphic to the facet of points with zero last coordinate in the $\mathbf{b}$-fiber of the extended matrix $(\mathcal{A}, \mathbf{b}) \in \mathbf{N}^{d \times (n+1)}$. Moreover, the $\mathbf{b}$-fiber of

$(\mathcal{A},\mathbf{b})$ is a Gröbner fiber since $x_{n+1} - x_1^{u_1}x_2^{u_2}\cdots x_n^{u_n}$ lies in $\mathcal{U}_{(\mathcal{A},\mathbf{b})}$ for every vertex $\mathbf{u} = (u_1,\ldots,u_n)$ of $conv(\pi^{-1}(\mathbf{b}))$. ■

We close with a geometric construction of the state polytope of a toric ideal.

Theorem 7.15. *The Minkowski sum of all Gröbner fibers is a state polytope for $I_\mathcal{A}$.*

Proof: Let $P = \sum_\mathbf{b} conv(\pi^{-1}(\mathbf{b}))$, where the sum is over all Gröbner degrees $\mathbf{b}$. Let ω and ω' be two generic vectors in $\mathbf{R}^n$. The following equivalences prove the claim:

$$\begin{aligned} & face_{-\omega}(P) = face_{-\omega'}(P) \\ \Longleftrightarrow\quad & face_{-\omega}(conv(\pi^{-1}(\mathbf{b}))) = face_{-\omega'}(conv(\pi^{-1}(\mathbf{b}))) \\ & \qquad\qquad\qquad\qquad\qquad \text{for every Gröbner degree } \mathbf{b} \\ \Longleftrightarrow\quad & in_\omega(I_\mathcal{A})_\mathbf{b} = in_{\omega'}(I_\mathcal{A})_\mathbf{b} \quad \text{for every Gröbner degree } \mathbf{b} \\ \Longleftrightarrow\quad & in_\omega(I_\mathcal{A}) = in_{\omega'}(I_\mathcal{A}). \end{aligned}$$

Here we use the abbreviation $in_{\omega'}(I_\mathcal{A})_\mathbf{b}$ to denote the k-linear span of $\{\mathbf{x^u} \in in_{\omega'}(I_\mathcal{A}) : \pi(\mathbf{u}) = \mathbf{b}\}$. ■

Corollary 7.16. *Let $\mathcal{B}$ be any finite subset of $\mathbf{N}\mathcal{A}$ which contains all Gröbner degrees. Then $\sum_{\mathbf{b}\in\mathcal{B}} conv(\pi^{-1}(\mathbf{b}))$ is a state polytope for $I_\mathcal{A}$.*

Proof: In the last equivalence of the proof of Theorem 7.15 we can replace "for all Gröbner degrees $\mathbf{b}$" by "for all $\mathbf{b} \in \mathcal{B}$". ■

Example 7.17. *(State polytope of the affine twisted cubic curve)*
Let $d = 1, n = 3$ and $\mathcal{A} = \{1,2,3\}$. The corresponding toric variety $X_\mathcal{A}$ is the twisted cubic curve in affine 3-space. The universal Gröbner basis of $I_\mathcal{A}$ is listed in the first row of Table 6-1. The Gröbner fibers are precisely the fibers over the scalars 2,3,4 and 6. For instance, $\pi^{-1}(4) = \{(4,0,0),(2,1,0),(0,2,0),(1,0,1)\}$ and $conv(\pi^{-1}(4))$ is a triangle (see Figure 7-2). By Theorem 7.15, the state polytope equals the following Minkowski sum:

$$State(I_\mathcal{A}) \quad = \quad conv(\pi^{-1}(2)) + conv(\pi^{-1}(3)) + conv(\pi^{-1}(4)) + conv(\pi^{-1}(6)).$$

The summands are one segment and three triangles. Their sum is a hexagon: This shows that $I_\mathcal{A} = \langle x^3 - z, x^2 - y\rangle$ has six distinct initial ideals. They are
(a) $\langle y, x^3\rangle$
(b) $\langle x^2, xy, y^2\rangle$
(c) $\langle x^2, xy, xz, y^3\rangle$
(d) $\langle x^2, xy, xz, z^2\rangle$
(e) $\langle x^2, z\rangle$
(f) $\langle y, z\rangle$
These ideals can be read off from Figure 7-2 as follows. Each vertex of the hexagon $State(I_\mathcal{A})$ is uniquely a sum of vertices of the fibers. For instance, the vertex labeled (a) is the sum of the upper-leftmost vertices x^2, z, xz and z^2. The initial ideal in question is generated by all monomials in Figure 7-2 not corresponding to the four vertices. ■

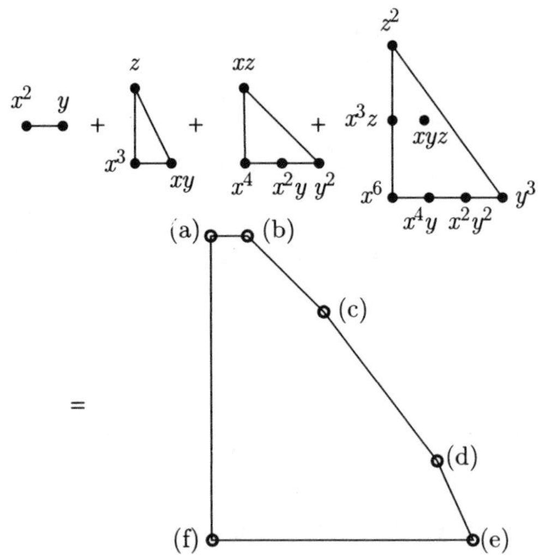

Figure 7-2. State polytope of the affine twisted cubic curve.

Exercises.

(1) Show that all Gröbner fibers of a Lawrence type matrix are one-dimensional.

(2) Show that the ratio of cardinalities $card(Gr_{\mathcal{A}})/card(\mathcal{U}_{\mathcal{A}})$ can be arbitrarily large.

(3) Suppose that F is the face of $P = conv(\pi^{-1}(\mathbf{b}))$ supported by $\omega \in \mathbf{R}^n$. Express the normal cone $\mathcal{N}_P(F)$ in terms of the Graver basis $Gr_{\mathcal{A}}$.

(4) Let $d = 3, n = 10$ and $\mathcal{A} = \{(i,j,k) \in \mathbf{N}^3 \; : \; i+j+k = 3\}$. Compute the Graver basis $Gr_{\mathcal{A}}$ and the universal Gröbner basis $\mathcal{U}_{\mathcal{A}}$. (The toric variety $\mathcal{V}(I_{\mathcal{A}})$ is the Veronese surface of degree nine in P^9.)

(5) Let $\mathcal{A}$ as in the previous exercise and $\mathcal{A}' := \mathcal{A} \setminus \{(1,1,1)\}$. Show that the toric ideal $I_{\mathcal{A}'}$ is generated by quadrics but it possesses no quadratic Gröbner basis.

(6) Consider all monomials $\mathbf{x}^{\mathbf{u}}$ with $\mathbf{u}$ not a vertex of its fiber $conv(\pi^{-1}(\pi(\mathbf{u})))$.
   (a) Show that they are the monomials in a monomial ideal $M_{\mathcal{A}}$.
   (a) Give an algorithm for computing $M_{\mathcal{A}}$.
   (b) Compute $M_{\mathcal{A}}$ for $d = 1, n = 4$, $\mathcal{A} = \{1,2,3,4\}$.

(7) Prove that the elements in the universal Gröbner basis $\mathcal{U}_{\mathcal{A}}$ are in bijection with the edge directions of the state polytope $State(I_{\mathcal{A}})$.

(8) Compute the state polytope of the ideal of 2×2-minors of a 2×4-matrix of indeterminates.

Notes:

The contents of this chapter is taken from (Sturmfels & Thomas 1994). A more general version of Theorem 7.8, valid for binomial ideals associated to arbitrary lattices, can be found in (Sturmfels, Weismantel & Ziegler 1994).

CHAPTER 8

Regular Triangulations

Let I be any ideal in $k[\mathbf{x}] = k[x_1, \ldots, x_n]$ and let $\prec$ be any term order. The passage from I to its initial ideal $in_\prec(I)$ is a flat deformation (Eisenbud 1995, Section 15.8). Here the zero set of I gets deformed into the zero set of the monomial ideal $in_\prec(I)$. The deformed zero set is a union of linear coordinate subspaces. It is convenient to identify the zero set of $in_\prec(I)$ with a simplicial complex.

The *initial complex* $\Delta_\prec(I)$ of I with respect to $\prec$ is the simplicial complex on the vertex set $\{1, 2, \ldots, n\}$ defined by the following rule. A subset $F \subset \{1, \ldots, n\}$ is a face of $\Delta_\prec(I)$ if there is no polynomial $f \in I$ whose initial monomial $in_\prec(f)$ has support F. Equivalently, $\Delta_\prec(I)$ is the simplicial complex whose *Stanley-Reisner ideal* is the radical of $in_\prec(I)$. As an example consider the principal ideal

$$I \quad = \quad \langle x_1 x_2 x_3 \; - \; x_4^3 \rangle .$$

The two initial complexes of I are given in Figure 8-1. The left picture corresponds to the choice of initial term $x_1 x_2 x_3$ while the right picture corresponds to x_4^3.

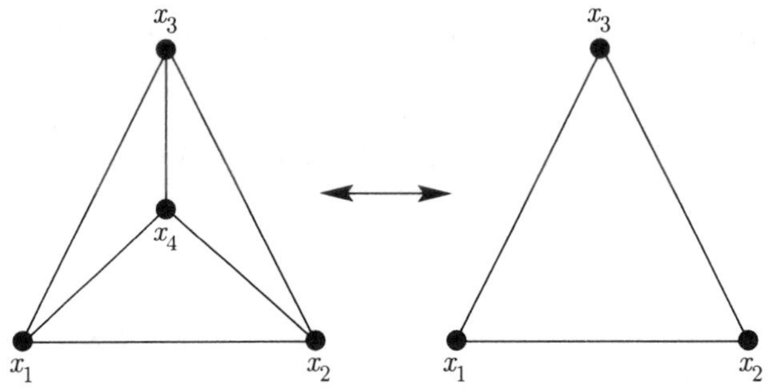

Figure 8-1. Two initial complexes of a toric surface.

It is our objective in this chapter to generalize the geometry in this little example. We shall determine the initial complexes of an arbitrary toric ideal $I_\mathcal{A}$.

If σ is a subset of $\mathcal{A}$ then we write $pos(\sigma)$ for the cone spanned by σ. A *triangulation* of $\mathcal{A}$ is a collection Δ of subsets of $\mathcal{A}$ such that $\{ pos(\sigma) : \sigma \in \Delta\}$ is the set of cones in a simplicial fan whose support equals $pos(\mathcal{A})$. We shall identify our given set of lattice points $\mathcal{A} = \{\mathbf{a}_1, \ldots, \mathbf{a}_n\} \subset \mathbf{Z}^d$ with the index set $\{1, \ldots, n\}$.

Example 8.1. *(Triangulations associated with the twisted cubic curve)*
Let $d = 2, n = 4$. The set $\mathcal{A} = \{(3,0),(2,1),(1,2),(0,3)\}$ has precisely four distinct triangulations: $\Delta^1 = \{\{1,2\},\{2,3\},\{3,4\}\}$, $\Delta^2 = \{\{1,2\},\{2,4\}\}$, $\Delta^3 = \{\{1,3\},\{3,4\}\}$, and $\Delta^4 = \{\{1,4\}\}$. ■

Every sufficiently generic vector $\omega \in \mathbf{R}^n$ defines a triangulation Δ_ω as follows: A subset $\{i_1, \ldots, i_r\}$ is a face of Δ_ω if there exists a vector $\mathbf{c} = (c_1, \ldots, c_d) \in \mathbf{R}^d$ such that

$$\begin{aligned} \mathbf{a}_j \cdot \mathbf{c} &= \omega_j \text{ if } j \in \{i_1, \ldots, i_r\} \text{ and} \\ \mathbf{a}_j \cdot \mathbf{c} &< \omega_j \text{ if } j \in \{1, \ldots, n\} \backslash \{i_1, \ldots, i_r\}. \end{aligned} \tag{8.1}$$

A triangulation Δ of $\mathcal{A}$ is called *regular* (or *coherent*) if $\Delta = \Delta_\omega$ for some $\omega \in \mathbf{R}^n$.

The regular triangulation Δ_ω can also be constructed geometrically:

(a) Using the coordinates of ω as "heights", we lift the configuration $\mathcal{A}$ into the next dimension. The result is the configuration $\widehat{\mathcal{A}} = \{(\mathbf{a}_1, \omega_1), \ldots, (\mathbf{a}_n, \omega_n)\} \subset \mathbf{R}^{d+1}$.

(b) The "lower faces" of the cone $pos(\widehat{\mathcal{A}})$ form a d-dimensional polyhedral complex. (A face is "lower" if it has a normal vector with negative last coordinate). The triangulation Δ_ω is the image of this complex under projection onto the first d coordinates.

Example 8.1. *(continued)* All four triangulations of $\mathcal{A}$ are regular:

$$\Delta^1 = \Delta_{(1,0,0,1)}, \quad \Delta^2 = \Delta_{(1,0,1,1)}, \quad \Delta^3 = \Delta_{(1,1,0,1)}, \quad \Delta^4 = \Delta_{(0,1,1,0)}.$$

Example 8.2. *(A non-regular triangulation)*
Let $d = 3$, $n = 6$ and $\mathcal{A} = \{(4,0,0), (0,4,0), (0,0,4), (2,1,1), (1,2,1), (1,1,2)\}$. The following collection of 3-sets defines a triangulation of $\mathcal{A}$. Notice the $\mathbf{Z}_3$-symmetry.

$$\Delta = \{\{1,2,5\}, \{1,3,4\}, \{1,4,5\}, \{2,3,6\}, \{2,5,6\}, \{3,4,6\}, \{4,5,6\}\}.$$

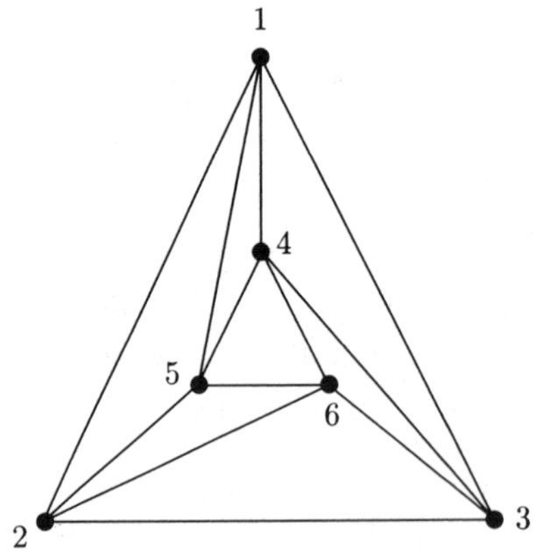

Figure 8-2. A non-regular triangulation.

We shall prove that Δ is not regular. Suppose that $\Delta = \Delta_\omega$ for some $\omega \in \mathbf{R}^6$. Then we can choose $\mathbf{c} \in \mathbf{R}^3$ to satisfy (8.1) for $\{1,2,5\}$. Consider the linear dependency

$$\mathbf{a}_1 - \mathbf{a}_2 - 4 \cdot \mathbf{a}_4 + 4 \cdot \mathbf{a}_5 = 0.$$

By taking the inner product of $\mathbf{c}$ with each term in this identity, we find that

$$\omega_1 - \omega_2 - 4\cdot\omega_4 + 4\cdot\omega_5 \quad < \quad 0. \tag{8.2}$$

The same reasoning applied to the equivalent triples $\{2,3,6\}$ and $\{3,1,4\}$ yields

$$\omega_2 - \omega_3 - 4\cdot\omega_5 + 4\cdot\omega_6 \; < \; 0 \qquad \text{and} \qquad \omega_3 - \omega_1 - 4\cdot\omega_6 + 4\cdot\omega_4 \; < \; 0. \tag{8.3}$$

The sum of the three inequalities in (8.2) and (8.3) equals $0 < 0$, a contradiction. ∎

Theorem 8.3. *The regular triangulations of $\mathcal{A}$ are the initial complexes of the toric ideal $I_\mathcal{A}$. More precisely, if $\omega \in \mathbf{R}^n$ represents $\prec$ for $I_\mathcal{A}$, then $\Delta_\prec(I_\mathcal{A}) = \Delta_\omega$.*

Proof: Let $\prec$ be any term order and suppose $\omega \in \mathbf{R}^n$ represents $\prec$ in the sense that $in_\prec(I_\mathcal{A}) = in_\omega(I_\mathcal{A})$. For any $\mathbf{b} \in \mathbf{R}^d$ consider the linear programming problem

$$\text{minimize } \mathbf{u}\cdot\omega \quad \text{subject to} \quad \mathbf{u} \in \mathbf{R}^n,\ \mathbf{u} \geq 0 \text{ and } u_1\mathbf{a}_1 + \cdots + u_n\mathbf{a}_n = \mathbf{b}. \tag{8.4}$$

The linear program dual to (8.4) takes the form

$$\text{maximize } \mathbf{c}\cdot\mathbf{b} \text{ subject to} \quad \mathbf{c} \in \mathbf{R}^d,\ \ \mathbf{a}_1\cdot\mathbf{c} \leq \omega_1,\ \ldots,\ \mathbf{a}_n\cdot\mathbf{c} \leq \omega_n. \tag{8.5}$$

For any subset F of $\{1,2,\ldots,n\}$ the following statements are equivalent:

F is a face of Δ_ω

$\Longleftrightarrow$ there exists a feasible solution $\mathbf{c}$ of (8.5) such that $F = \{j \in \{1,\ldots,n\} : \mathbf{a}_j\cdot\mathbf{c} = \omega_j\}$

$\Longleftrightarrow$ $\exists\, \mathbf{b} \in \mathbf{Z}^d$: an optimal solution $\mathbf{c}$ of (8.5) satisfies $F = \{j : \mathbf{a}_j\cdot\mathbf{c} = \omega_j\}$

$\Longleftrightarrow$ $\exists\, \mathbf{b} \in \mathbf{Z}^d$: an optimal solution $\mathbf{u}$ of (8.4) satisfies $supp(\mathbf{u}) = F$

$\Longleftrightarrow$ $\exists\, \mathbf{b} \in \mathbf{Z}^d$: an optimal solution $\mathbf{u}$ of (8.4) satisfies $supp(\mathbf{u}) = F$ and is integral

$\Longleftrightarrow$ there exists $\mathbf{x}^\mathbf{u}$ such that $F = supp(\mathbf{x}^\mathbf{u})$ and every power of $\mathbf{x}^\mathbf{u}$ is standard

$\Longleftrightarrow$ F is a face of $\Delta_\prec(I_\mathcal{A})$.

The first and last equivalences in the chain above are translations of the definitions of Δ_ω and $\Delta_\prec(I_\mathcal{A})$ respectively. The second equivalence holds because every point $\mathbf{c}$ in the feasible polyhedron of (8.5) lies in the relative interior of some face. The integer vector $\mathbf{b}$ is chosen to be a support vector of that face. The third equivalence holds because of *complementary slackness*; see §7.9 in (Schrijver 1986). For the fourth equivalence replace $\mathbf{b}$ by a suitable integer multiple of $\mathbf{b}$. In the fifth equivalence we are using the fact that every power of $\mathbf{x}^\mathbf{u}$ is standard if and only if every integer multiple of $\mathbf{u}$ solves its corresponding integer program (by Algorithm 5.6). The latter condition means that $\mathbf{u}$ is integral and solves its linear program (8.4). This completes the proof of Theorem 8.3. ∎

Corollary 8.4. *For generic $\omega \in \mathbf{R}^n$, the radical of the initial ideal of $I_\mathcal{A}$ equals*

$$\begin{aligned} Rad(in_\omega(I_\mathcal{A})) &= \langle x_{i_1}x_{i_2}\cdots x_{i_s} : \{i_1,i_2,\ldots,i_s\} \text{ is a minimal non-face of } \Delta_\omega \rangle \\ &= \bigcap_{\sigma\in\Delta_\omega} \langle x_i : i \notin \sigma \rangle. \end{aligned}$$

Proof: The first equality follows from Theorem 8.3. The second equality is a general formula for the prime decomposition of any square-free monomial ideal. ∎

In the remainder of this chapter we assume that the set $\mathcal{A}$ spans an affine hyperplane. By Lemma 4.14 this means that $I_{\mathcal{A}}$ is a homogeneous ideal of Krull dimension d. Each d-subset of $\{1, \ldots, n\}$ corresponds to a $(d-1)$-simplex in the affine span of $\mathcal{A}$. We can thus consider Δ_ω as a regular triangulation of the polytope $Q = conv(\mathcal{A})$ with vertices in $\mathcal{A}$. In particular, Theorem 8.3 and Corollary 8.4 give rise to an algebraic algorithm for computing triangulations of polytopes. We present some examples of polytopes of dimension $d - 1 = 3$.

Example 8.5. *(Lexicographic triangulations of some 3-polytopes)*
In each of the following cases we present the reduced Gröbner basis $\mathcal{G}$ with respect to the lexicographic term order defined by $x_1 \succ x_2 \succ \cdots \succ x_n$. The underlined initial terms define the minimal non-faces of a regular triangulation $\Delta_\prec$ of the polytope $Q = conv(\mathcal{A})$.

(a) *The regular octahedron:* $n = 6$, $Vol(Q) = deg(I_{\mathcal{A}}) = 4$.

$$\begin{aligned} \mathcal{A} &= \{(1,1,0,0),(1,0,1,0),(1,0,0,1),(0,1,1,0),(0,1,0,1),(0,0,1,1)\}, \\ \mathcal{G} &= \{\underline{x_1x_6} - x_3x_4,\ \underline{x_2x_5} - x_3x_4\}, \\ \Delta_\prec &= \{\{1,2,3,4\},\ \{1,5,3,4\},\ \{2,6,3,4\},\ \{5,6,3,4\}\}. \end{aligned}$$

(b) *The cyclic polytope with $n = 5$ vertices:* $Vol(Q) = deg(I_{\mathcal{A}}) = 8$.

$$\begin{aligned} \mathcal{A} &= \{1\} \times \{(0,0,0),(1,1,1),(2,4,8),(3,9,27),(4,16,64)\} \\ \mathcal{G} &= \{\underline{x_1x_3^6x_5} - x_2^4x_4^4\}. \end{aligned}$$

(c) *The regular 3-cube:* $n = 8$, $Vol(Q) = deg(I_{\mathcal{A}}) = 6 = \#$ tetrahedra in $\Delta_\prec$.

$$\begin{aligned} \mathcal{A} = &\{1\} \times \{(0,0,0),(0,0,1),(0,1,0),(0,1,1), \\ &\qquad (1,0,0),(1,0,1),(1,1,0),(1,1,1)\} \\ \mathcal{G} = &\{\underline{x_1x_4} - x_2x_3,\ \underline{x_1x_6} - x_2x_5,\ \underline{x_1x_7} - x_3x_5,\ \underline{x_1x_8} - x_4x_5,\ \underline{x_2x_7} - x_4x_5, \\ &\quad \underline{x_2x_8} - x_4x_6,\ \underline{x_3x_6} - x_4x_5,\ \underline{x_3x_8} - x_4x_7,\ \underline{x_5x_8} - x_6x_7\}. \end{aligned}$$

The triangulation $\Delta_\prec$ is given in the diagram below. Note that the minimal non-faces of $\Delta_\prec$ are precisely the nine underlined monomials above.

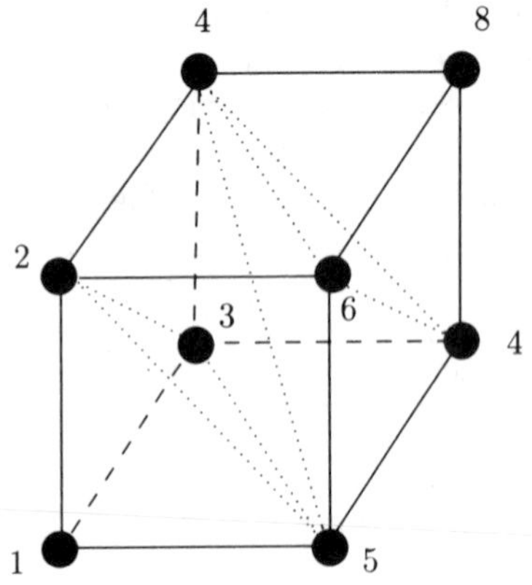

Figure 8-3. Lexicographic triangulation of the 3-cube.

(d) *The permutohedron* (a.k.a. *the truncated octahedron):*
Here $n = 24$ and $\mathcal{A} = \{ (\pi_1, \pi_2, \pi_3, \pi_4) : \pi$ permutation of $\{1,2,3,4\}\}$. The polytope $Q = conv(\mathcal{A})$ has normalized volume 96. The ideal $I_\mathcal{A}$ is minimally generated by 361 binomials, among which 99 have degree two and 262 have degree three. The reduced Gröbner basis $\mathcal{G}$ has 578 elements with maximum occurring degree six. Replacing the initial ideal by its radical, and computing its prime decomposition as in Corollary 8.4, we see that the regular triangulation $\Delta_\prec$ consists of precisely 42 tetrahedra. ∎

The two most popular term orders for Gröbner bases are the lexicographic order $\prec_{lex}$ and the reverse lexicographic order $\prec_{revlex}$. It comes as no surprise that the corresponding triangulations are the two most popular triangulations of a polytope. The *lexicographic triangulation* $\Delta_{\prec_{lex}}$ is known in combinatorics as the *placing triangulation*, and the *reverse lexicographic triangulation* $\Delta_{\prec_{revlex}}$ is known as the *pulling triangulation*. This correspondence was established in (Sturmfels 1991). For a nice geometric discussion of these triangulations see (Lee 1991).

We shall give recursive description of the two triangulations $\Delta_{lex}(\mathcal{A})$ and $\Delta_{revlex}(\mathcal{A})$, where $\mathcal{A} = \{\mathbf{a}_1 > \mathbf{a}_2 > \cdots > \mathbf{a}_n\}$ is any totally ordered set of points in $\mathbf{R}^d$. By a *face* (resp. *facet*) of $\mathcal{A}$ we mean a totally ordered subset of the form $F \cap \mathcal{A}$ where F is a face (resp. facet) of the polytope $conv(\mathcal{A})$. The two triangulations below are subsets of $2^\mathcal{A}$, the power set of $\mathcal{A}$, which are closed under taking subsets. In our formulas we tacitly assume that the closure operation of adding subsets is applied whenever needed to form a simplicial complex.

Proposition 8.6. *If $\mathcal{A}$ is affinely independent, then $\Delta_{lex}(\mathcal{A}) = \Delta_{revlex}(\mathcal{A}) = \{\mathcal{A}\}$. Else:*

$$\Delta_{revlex}(\mathcal{A}) \quad = \quad \bigcup_F \{ \{\mathbf{a}_n\} \cup G \; : \; G \in \Delta_{revlex}(F) \}, \tag{8.6}$$

where the union is over all facets F of $\mathcal{A}$ not containing $\mathbf{a}_n$ and

$$\Delta_{lex}(\mathcal{A}) = \Delta_{lex}(\mathcal{A}\backslash\{\mathbf{a}_1\}) \bigcup \{ \{\mathbf{a}_1\} \cup G \; : \; G \text{ is face of } \Delta_{lex}(\mathcal{A}\backslash\{\mathbf{a}_1\}) \text{ which is visible from } \mathbf{a}_1 \}. \tag{8.7}$$

Proof: A subset F of $\mathcal{A}$ is a face of Δ_{lex} if and only if there exists a monomial $\mathbf{x}^\mathbf{u}$ which has support F and is not nilpotent modulo $in_{lex}(I_\mathcal{A})$, and similarly for Δ_{revlex}. We shall proceed by induction on $n = card(\mathcal{A})$. The case where $\mathcal{A}$ is affinely independent is obvious.

First consider the lexicographic term order. It has the property

$$in_{lex}(I_\mathcal{A}) \cap k[x_2, \ldots, x_n] \quad = \quad in_{lex}(I_{\mathcal{A}\backslash\{\mathbf{a}_1\}}).$$

Equivalently, the standard monomial modulo $I_\mathcal{A}$ not containing x_1 are precisely the standard monomials modulo $I_{\mathcal{A}\backslash\{\mathbf{a}_1\}}$. This implies that the faces of $\Delta_{lex}(\mathcal{A})$ not containing $\mathbf{a}_1$ are precisely the faces of $\Delta_{lex}(\mathcal{A}\backslash\{\mathbf{a}_1\})$. The right hand union in (8.7) is the unique completion of $\Delta_{lex}(\mathcal{A}\backslash\{\mathbf{a}_1\})$ to a triangulation of $\mathcal{A}$. It must therefore be equal to $\Delta_{lex}(\mathcal{A})$.

Consider the reverse lexicographic term order. Let $\mathbf{x}^\mathbf{u}$ be a monomial which is not nilpotent modulo $in_{revlex}(I_\mathcal{A})$ and $H = supp(\mathbf{x}^\mathbf{u})$. We must show that H is a subset of one of the sets in the union in (8.6). First suppose that x_n does not appear in $\mathbf{x^u}$. By the properties of reverse lexicographic order, no monomial in the same fiber as $\mathbf{x^u}$ contains the variable x_n. This means that H is contained in a proper face of $\mathcal{A}$, and hence $H \subseteq F$ for some facet F of $\mathcal{A}$ not containing $\mathbf{a}_n$. By the induction hypothesis, we have $H \in \Delta_{revlex}(F)$. Next suppose that x_n appears in $\mathbf{x^u}$ with degree $i \geq 1$. Then the support of $\mathbf{x^u}/x_n^i$ lies in $\Delta_{revlex}(F)$ for some F as above. Hence $H = \{\mathbf{a}_n\} \cup supp(\mathbf{x^u}/x_n^i)$ appears in the union on the right hand side of (8.6). ∎

Example 8.7. *(Lexicographic versus reverse lexicographic triangulations)* Consider the cubic Veronese surface in P^9. The set $\mathcal{A} = \{(i_1, i_2, i_3) \in \mathbf{N}^3 : i_1 + i_2 + i_3 = 3\}$ is labeled as indicated below, and the variables are ordered $x_1 \succ x_2 \succ \cdots \succ x_{10}$.

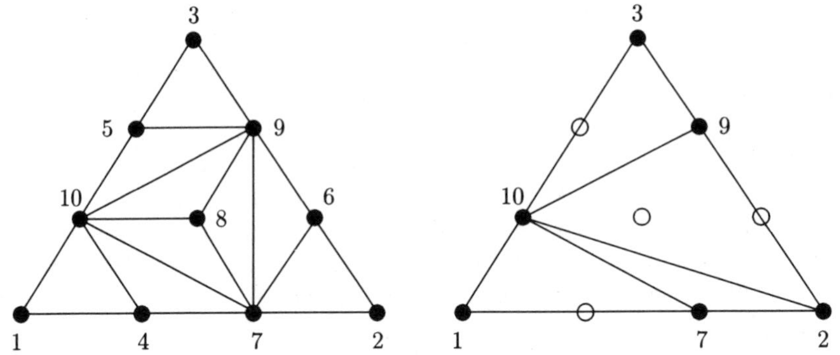

Figure 8-4. Lexicographic and reverse lexicographic triangulation.

The lexicographic triangulation is shown on the left and the reverse lexicographic triangulation is shown on the right. From these pictures we can determine the radicals of the initial ideals:

$$\begin{aligned} Rad\big(in_{lex}(I_\mathcal{A})\big) &= \langle x_1x_2, x_1x_3, x_1x_5, x_1x_6, x_1x_7, x_1x_8, x_1x_9, x_2x_{10}, x_2x_3, x_2x_4, \\ &\quad x_2x_5, x_2x_8, x_2x_9, x_3x_{10}, x_3x_4, x_3x_6, x_3x_7, x_3x_8, x_4x_5, x_4x_6, \\ &\quad x_4x_8, x_4x_9, x_5x_6, x_5x_7, x_5x_8, x_6x_{10}, x_6x_8, x_7x_9x_{10}\rangle \quad \textit{and} \\ Rad\big(in_{revlex}(I_\mathcal{A})\big) &= \langle x_1x_2, x_1x_3, x_1x_9, x_2x_3, x_3x_7, x_4, x_5, x_6, x_7x_9, x_8\rangle. \end{aligned}$$

The generators of these two ideals are the minimal non-faces in the two triangulations. Note that the left triangulation is unimodular, i.e., every triangle has unit area. By Corollary 8.9 this implies that $in_{lex}(I_\mathcal{A})$ is equal to its radical listed above. ∎

Let $I_\mathcal{A}$ be a homogeneous toric ideal. The degree of any initial ideal $in_\prec(I_\mathcal{A})$ is equal to the normalized volume $Vol(Q)$ of the polytope $Q = conv(\mathcal{A})$. This follows from Theorem 4.16 and the fact that the degree of a homogeneous ideal is preserved under passing to the initial ideal. We shall prove the following stronger result. It implies Theorem 4.16 because the volume of Q equals the sum of the volumes of the $(d-1)$-simplices in any triangulation of Q, and the degree of a monomial ideal is the sum of the multiplicities of all associated primes of maximal dimension.

Theorem 8.8. *Let σ be any $(d-1)$-simplex in the regular triangulation $\Delta_{\prec}(I_{\mathcal{A}})$. The normalized volume of σ equals the multiplicity of the prime ideal $\langle\, x_i \,:\, i \notin \sigma \rangle$ in $in_{\prec}(I_{\mathcal{A}})$.*

Proof: The multiplicity M of $\langle x_i : i \notin \sigma \rangle$ in $in_{\prec}(I_{\mathcal{A}})$ can be computed as follows. Let $\phi : k[\mathbf{x}] \to k[\, x_i : i \notin \sigma]$ be the homomorphism which maps x_j to 1 for $j \in \sigma$ and x_i to x_i for $i \notin \sigma$. Then $\phi(in_{\prec}(I_{\mathcal{A}}))$ is a zero-dimensional ideal and M equals the k-dimension of $k[x_i : i \notin \sigma]/\phi(in_{\prec}(I_{\mathcal{A}}))$. It follows from (8.1) that $\prec$ can be represented by a weight vector $\omega \in \mathbf{R}^n$ such that $\omega_j = 0$ for $j \in \sigma$ and $\omega_i > 0$ for $i \notin \sigma$. This representation implies

$$\phi(in_{\prec}(I_{\mathcal{A}})) \quad = \quad in_{\omega}(\phi(I_{\mathcal{A}})). \tag{8.8}$$

Hence M is the k-dimension of $k[\, x_i \,:\, i \notin \sigma]/\phi(I_{\mathcal{A}})$. Let $\mathcal{L}$ denote the image of $ker(\mathcal{A})$ under the homomorphism $\mathbf{Z}^n \to \mathbf{Z}^{n-card(\sigma)}$ which deletes all coordinates in σ. Then

$$\phi(I_{\mathcal{A}}) \quad = \quad \langle\, \mathbf{x}^{\mathbf{u}^+} - \mathbf{x}^{\mathbf{u}^-} \,:\, \mathbf{u} = \mathbf{u}^+ - \mathbf{u}^- \in \mathcal{L}\,\rangle. \tag{8.9}$$

To compute the normalized volume of the $(d-1)$-simplex σ, we shall assume that the $d \times n$-matrix $\mathcal{A}$ has rank d and that the $d \times d$-minors of $\mathcal{A}$ are relatively prime. Then $Vol(\sigma) = \pm det(\mathcal{A}_\sigma)$, the determinant of the $d \times d$-minor with column indices σ. Let $\mathcal{B}$ be any integer $(n-d) \times n$-matrix whose rows constitute a $\mathbf{Z}$-basis for $ker(\mathcal{A})$. It is a basic fact of linear algebra that complementary minors of $\mathcal{A}$ and $\mathcal{B}$ are equal up to a sign:

$$Vol(\sigma) \quad = \quad |\,det(\mathcal{A}_\sigma)| \quad = \quad |\,det(\mathcal{B}_{\{1,\ldots,n\}\backslash\sigma})|. \tag{8.10}$$

The $(n-d) \times (n-d)$-minor on the right hand side equals the index of the sublattice $\mathcal{L}$ in $\mathbf{Z}^{n-card(\sigma)}$. It is to be shown in Exercise (1) below that $[\mathbf{Z}^{n-card(\sigma)} : \mathcal{L}]$ equals the k-dimension of $k[x_j \,:\, j \notin \sigma]$ modulo the ideal in (8.9). This completes the proof. ∎

A triangulation Δ of $\mathcal{A}$ is *unimodular* if $Vol(\sigma) = 1$ for every maximal simplex σ in Δ.

Corollary 8.9. *The initial ideal $in_{\prec}(I_{\mathcal{A}})$ is square-free if and only if the corresponding regular triangulation $\Delta_{\prec}$ of $\mathcal{A}$ is unimodular.*

Proof: The only-if-direction follows directly from Theorem 8.8. To prove the if-direction, suppose that $\Delta_{\prec}$ is unimodular but $in_{\prec}(I_{\mathcal{A}})$ is not square-free. Choose a standard monomial $\mathbf{x}^{\mathbf{u}}$ such that $\mathbf{x}^{2\mathbf{u}}$ is not standard. Let $\mathbf{b} = \pi(\mathbf{u})$ and let σ be a $(d-1)$-simplex in $\Delta_{\prec}$ whose positive span in $\mathbf{R}^d$ contains $\mathbf{b}$. Represent $\prec$ by a weight vector $\omega \in \mathbf{R}^n$ such that $\omega_j = 0$ for $j \in \sigma$ and $\omega_i > 0$ for $i \notin \sigma$. The linear system of equations

$$\sum_{i \in \sigma} \lambda_i \mathbf{a}_i \quad = \quad \mathbf{b}$$

has a unique rational solution $(\lambda_i \,:\, i \in \sigma)$. Our choice of σ means that $\lambda_i \geq 0$ for all $i \in \sigma$. Moreover, all λ_i are integers because $Vol(\sigma) = [\mathbf{Z}\mathcal{A} : \mathbf{Z}\{\mathbf{a}_i \,:\, i \in \sigma\}]$ is equal to one. The monomial $\prod_{i \in \sigma} x_i^{\lambda_i}$ has ω-weight zero, and it is congruent modulo $I_{\mathcal{A}}$ to the standard monomial $\mathbf{x}^{\mathbf{u}}$. Therefore $\mathbf{x}^{\mathbf{u}} = \prod_{i \in \sigma} x_i^{\lambda_i}$, and hence $\mathbf{x}^{2\mathbf{u}} = \prod_{i \in \sigma} x_i^{2\lambda_i} \notin Rad(in_{\prec}(I_{\mathcal{A}}))$. This is a contradiction to our assumption that $\mathbf{x}^{2\mathbf{u}}$ is a non-standard monomial. ∎

The matrix $\mathcal{A}$ is called *unimodular* if all triangulations of $\mathcal{A}$ are unimodular. If $rank(\mathcal{A}) = d$, then $\mathcal{A}$ is unimodular if and only if all non-zero $d \times d$-minors of $\mathcal{A}$ have the same absolute value. See (Schrijver 1986; Theorem 19.2) for another characterization.

Remark 8.10. *A matrix $\mathcal{A}$ is unimodular if and only if all initial ideals of the toric ideal $I_{\mathcal{A}}$ are square-free.*

Proof: This follows directly from Corollary 8.9. ∎

Unimodular matrices have the following important property.

Proposition 8.11. *If $\mathcal{A}$ is a unimodular matrix, then the set of circuits $\mathcal{C}_{\mathcal{A}}$ equals the Graver basis $Gr_{\mathcal{A}}$.*

Proof: By Lemma 4.9 and unimodularity, every circuit $\mathbf{v}$ of $\mathcal{A}$ has its coordinates in $\{0, -1, +1\}$. Let $\mathbf{x}^{\mathbf{u}^+} - \mathbf{x}^{\mathbf{u}^-}$ be any element in the Graver basis of $I_{\mathcal{A}}$. By Lemma 4.10, there exists a circuit $\mathbf{x}^{\mathbf{v}^+} - \mathbf{x}^{\mathbf{v}^-}$ such that $supp(\mathbf{v}^+) \subseteq supp(\mathbf{u}^+)$ and $supp(\mathbf{v}^-) \subseteq supp(\mathbf{u}^-)$. The monomials $\mathbf{x}^{\mathbf{v}+}$ and $\mathbf{x}^{\mathbf{v}^-}$ are square-free, by the remark above. This implies that $\mathbf{x}^{\mathbf{v}^+}$ divides $\mathbf{x}^{\mathbf{u}^+}$ and $\mathbf{x}^{\mathbf{v}^-}$ divides $\mathbf{x}^{\mathbf{u}^-}$. Since $\mathbf{u}$ was assumed to lie in $Gr_{\mathcal{A}}$, this implies $\mathbf{u} = \mathbf{v} \in \mathcal{C}_{\mathcal{A}}$. The reverse inclusion $\mathcal{C}_{\mathcal{A}} \subseteq Gr_{\mathcal{A}}$ was established in Proposition 4.11. ∎

A prototypical example of a unimodular matrix is the $(s+t) \times (st)$-matrix in (5.1). The corresponding polytope $Q = conv(\mathcal{A})$ equals the product of simplices $\Delta_{s-1} \times \Delta_{t-1}$. The reader is asked in Exercise (9) below to prove that this configuration is unimodular.

Example 8.12. *(The staircase triangulation of the product of two simplices)* The set of variables is $X = \{\, x_{ij} : 1 \leq i \leq s, 1 \leq j \leq t\}$. We define a partial order on X by setting $x_{ik} \leq x_{jl}$ whenever $i \leq j$ and $k \leq l$. Consider the reduced Gröbner basis $\mathcal{G}_{\prec}$ given in Proposition 5.4. The underlined initial monomials in (5.7) are precisely the incomparable pairs in our poset, and by Theorem 8.3, they define the minimal non-faces in a triangulation $\Delta_{\prec}$ of $\Delta_{s-1} \times \Delta_{t-1}$. The maximal simplices in $\Delta_{\prec}$ are the maximal chains in the poset. In the matrix they form the "staircases" from the upper left corner x_{11} to the lower right corner x_{st}. Therefore $\Delta_{\prec}$ is called the *staircase triangulation* of $\Delta_{s-1} \times \Delta_{t-1}$. We note that the staircase triangulation is both lexicographic and reverse lexicographic with respect to the usual row-wise variable ordering $x_{11} \succ x_{12} \succ \cdots \succ x_{st}$. ∎

Example 8.13. *(A non-regular triangulation of the product of two tetrahedra)* For a long time it was unknown whether all triangulations of a product of simplices are regular. The following solution to this problem for $s = t = 4$ was given by De Loera (1995). To simplify the presentation, we write the 4×4-matrix of indeterminates as

$$(x_{ij}) \quad = \quad \begin{pmatrix} a & b & c & d \\ e & f & g & h \\ i & j & k & l \\ m & n & o & p \end{pmatrix}$$

Consider the following square-free monomial ideal:

$$J \quad = \quad \langle af,\, \underline{ag},\, ah,\, al,\, an,\, ao,\, ap,\, \underline{bi},\, bl,\, bp,\, cf,\, ch,\, ci,\, cj,\, cl,\, \underline{cn},\, cp,\, df,\, dgj,$$
$$\underline{el},\, en,\, eo,\, ep,\, fi,\, fk,\, fl,\, fo,\, fp,\, gi,\, gl,\, \underline{ho},\, in,\, io,\, ip,\, jo,\, \underline{jp},\, lo \rangle.$$

Its associated simplicial complex $\Delta(J)$ is a triangulation of $\Delta_3 \times \Delta_3$. One verifies this by checking that the ideal J has the same Hilbert series in the fine $\mathbf{N}^4 \times \mathbf{N}^4$-grading as the staircase initial ideal $in_{\prec}(I_{\mathcal{A}})$ in Example 8.12 (see Exercise (2) below). If the triangulation $\Delta(J)$ were regular, then there would exist weights $\omega = (A, B, C, ..., P) \in \mathbf{R}^{4\times 4}$ such that $J = in_\omega(I_{\mathcal{A}})$. The underlined generators of J impose the following six inequalities on the weights:

$$
\begin{aligned}
A+G &> C+E \\
B+I &> A+J \\
C+N &> B+O \\
E+L &> H+I \\
H+O &> G+P \\
J+P &> L+N
\end{aligned}
$$

These inequalities are inconsistent, because the sum of the right hand sides equals the sum of the left hand sides. This proves that the triangulation $\Delta(J)$ is not regular. ∎

With every regular triangulation Δ of a set $\mathcal{A}$ we associate the polyhedral cone

$$\mathcal{C}_\Delta \quad := \quad \{\omega \in \mathbf{R}^n \ : \ \Delta_\omega = \Delta\}. \tag{8.11}$$

The cone $\mathcal{C}_\Delta$ consists of all lifting functions which induce the triangulation Δ. The collection of these cones together with their faces is a polyhedral fan. We state without proof the following theorem due to Gel'fand, Kapranov and Zelevinsky (1994).

Theorem 8.14. *There exists an $(n-d)$-dimensional polytope $\Sigma(\mathcal{A})$, such that the normal cones at the vertices of $\Sigma(\mathcal{A})$ are precisely the cones $\mathcal{C}_\Delta$.*

The polytope $\Sigma(\mathcal{A})$ is called the *secondary polytope*, and its normal fan

$$\mathcal{N}(\Sigma(\mathcal{A})) \;=\; \{\mathcal{C}_\Delta : \Delta \,\text{regular triangulation of }\, \mathcal{A}\}$$

is called the *secondary fan.*

Proposition 8.15.

(a) The Gröbner fan of the toric ideal $I_{\mathcal{A}}$ is a refinement of the secondary fan $\mathcal{N}(\Sigma(\mathcal{A}))$. If $\mathcal{A}$ is unimodular, then the two fans coincide.

(b) The secondary polytope $\Sigma(\mathcal{A})$ is a Minkowski summand of the state polytope $State(I_{\mathcal{A}})$. If $\mathcal{A}$ is unimodular, then the two polytopes coincide.

Proof: It suffices to prove statement (a) since (b) is just a reformulation. Theorem 8.3 tells us that $in_\omega(I_{\mathcal{A}}) = in_{\omega'}(I_{\mathcal{A}})$ implies $\Delta_\omega = \Delta_{\omega'}$. This shows that each Gröbner cone lies entirely in a secondary cone $\mathcal{C}_\omega$. The reverse implication holds if $\mathcal{A}$ is unimodular, by Remark 8.10. ∎

Example 8.16. It is possible that the state polytope equals the secondary polytope, even if $\mathcal{A}$ is not unimodular. An example with this property (for $d = 6, n = 8$) is the Lawrence lifting $\mathcal{A} = \Lambda(\mathcal{A}')$ of

$$\mathcal{A}' \quad = \quad \begin{pmatrix} 1 & -2 & 1 & 0 \\ 1 & -1 & 0 & 1 \end{pmatrix}.$$

The matrix $\mathcal{A}'$ is not unimodular, and this implies that $\mathcal{A} = \Lambda(\mathcal{A}')$ is not unimodular. An explicit computation shows that $State(I_\mathcal{A}) = \Sigma(\mathcal{A})$ is a planar octagon. ■

Exercises.

(1) Let $\mathcal{L}$ be an n-dimensional sublattice of $\mathbf{Z}^n$ and consider the ideal

$$I_\mathcal{L} \quad := \quad \langle \mathbf{x}^{\mathbf{u}^+} - \mathbf{x}^{\mathbf{u}^-} \; : \mathbf{u} \in \mathcal{L} \rangle.$$

Show that $k[\mathbf{x}]/I_\mathcal{L}$ is a k-vector space of dimension $[\mathbf{Z}^n : \mathcal{L}]$.

(2) Let $J = \langle \mathbf{x}^{\mathbf{u}_1}, \ldots, \mathbf{x}^{\mathbf{u}_s} \rangle$ be a *square-free* monomial ideal in $k[\mathbf{x}]$, and let $\hat{\pi}$ as in (4.1). Its Hilbert function in the d-variate grading defined by $deg(x_i) = \mathbf{a}_i$ is the rational generating function

$$H(J; \mathbf{t}) \quad = \quad \frac{\sum_{\nu \subseteq \{1,\ldots,s\}} (-1)^{|\nu|} \cdot \hat{\pi}\big(lcm(\{\mathbf{x}^{\mathbf{u}_j} \; : \; j \in \nu\})\big)}{\prod_{i=1}^n (1 - \mathbf{t}^{\mathbf{a}_i})}.$$

Here "lcm" denotes the least common multiple of a set of monomials in $k[\mathbf{x}]$. Show that the simplicial complex defined by J is a triangulation of $\mathcal{A}$ if and only if

$$H(J; \mathbf{t}) \quad = \quad \sum \{ \mathbf{t}^\mathbf{b} \; : \; \mathbf{b} \in \mathbf{N}\mathcal{A} \}.$$

(3) Let $\mathcal{A}$ be the set of vertices of a planar n-gon in convex position. Show that every triangulation of $\mathcal{A}$ is lexicographic (and hence regular). Give an example of a triangulation of a hexagon which is not reverse lexicographic.

(4) Find a regular triangulation that is neither lexicographic nor reverse lexicographic.

(5) Prove: If the secondary polytope of $\mathcal{A}$ equals the state polytope of $I_\mathcal{A}$, then the set of circuits $\mathcal{C}_\mathcal{A}$ equals the universal Gröbner basis $\mathcal{U}_\mathcal{A}$. Does the converse hold ?

(6) Compute all eight initial ideals of the toric ideal in Example 8.16.

(7) Compute the secondary polytope and the state polytope for the twisted cubic curve in Example 8.1.

(8) Let I be an arbitrary ideal in $k[\mathbf{x}]$ and $\Delta_\prec(I)$ its initial complex with respect to any term order $\prec$. Show that the Krull dimension of $k[\mathbf{x}]/I$ is equal to $dim\big(\Delta_\prec(I)\big) + 1$.

(9) Show that the set $\mathcal{A} = \{\mathbf{e}_i - \mathbf{e}_j \; : \; 1 \leq i < j \leq d\}$ is unimodular (cf. Exercise (8) in Chapter 4). Deduce that the product of simplices $\Delta_{s-1} \times \Delta_{t-1}$ is unimodular. (Hint: Consider the complete bipartite graph $K_{s,t}$ as a subgraph of the complete graph K_{s+t}.)

Notes:

Initial complexes were studied in (Kalkbrener & Sturmfels 1995) for general prime ideals. The connection between regular triangulations and initial ideals of toric ideals was introduced in (Sturmfels 1991). An extension of Theorem 8.3 to ideals of lattices can be found in (Sturmfels, Weismantel & Ziegler 1995). Theorem 8.8 was proved in (Kapranov, Sturmfels & Zelevinsky 1992). Connections to sparse elimination theory are also explored in (Sturmfels 1993b). The construction of the secondary polytope $\Sigma(\mathcal{A})$ is due to Gel'fand, Kapranov & Zelevinsky (1994). An alternative proof of Theorem 8.14 appears in (Billera, Sturmfels & Filliman 1990).

CHAPTER 9

The Second Hypersimplex

We now apply our theory to a specific family of polytopes. Let $\mathcal{A}_d = \{\mathbf{e}_i + \mathbf{e}_j : 1 \leq i < j \leq d\}$. This is the set of column vectors of the vertex-edge incidence matrix of the complete graph K_d. The convex hull of $\mathcal{A}_d$ is called the *second hypersimplex* of order d and is denoted $\Delta(2,d)$. The second hypersimplex is a $(d-1)$-dimensional polytope in $\mathbf{R}^d$, with $n = \binom{d}{2}$ vertices. Its toric ideal $I_{\mathcal{A}_d}$ is the kernel of the map

$$\Phi \;:\; k[\,x_{ij} : 1 \leq i < j \leq d\,] \;\rightarrow\; k[t_1,\ldots,t_d]\,,\;\; x_{ij} \;\mapsto\; t_i t_j. \tag{9.1}$$

The variables x_{ij} are indexed by the edges in the complete graph K_d.

This chapter is organized as follows. We first describe a quadratic, square-free Gröbner basis for $I_{\mathcal{A}_d}$ and the associated regular triangulation of $\Delta(2,d)$ into unit simplices. We then compare the state polytope of $I_{\mathcal{A}_d}$ with the secondary polytope of $\mathcal{A}_d$, and we describe the universal Gröbner basis of $\mathcal{A}_d$ for $d \leq 8$. Finally, we shall discuss the integer programming problem and the sampling problem associated with $\mathcal{A}_d$.

We identify the vertices of K_d with the vertices of a regular d-gon in the plane labeled clockwise from 1 to d. Between any two vertices of K_d there are two paths that use only edges of the d-gon. We define the *circular distance* between two vertices to be the length of the shorter path. For example, vertices 1 and d are at circular distance 1 inside K_d and the pair $\{1,6\}$ has distance 3 inside K_8.

In what follows we use the term *edge* for the closed line segment joining any two vertices in the convex d-gon. We define the *weight* of the variable x_{ij} as the number of edges of K_d which do not meet the edge (i,j). For instance, if $d=5$, then the variables $x_{12}, x_{23}, x_{34}, x_{45}, x_{15}$ have weight 3, and the variables $x_{13}, x_{24}, x_{35}, x_{14}, x_{25}$ have weight 1. Let $\succ$ denote any term order that refines the partial order on monomials specified by these weights. Given any pair of non-intersecting edges $(i,j),(k,l)$ of K_d, one of the pairs $(i,k),(j,l)$ or $(i,l),(j,k)$ meets in a point. With the disjoint edges $(i,j),(k,l)$, we associate the binomial $x_{ij}x_{kl} - x_{il}x_{jk}$ where $(i,l),(j,k)$ is the intersecting pair. We denote by $\mathcal{C}$ the set of all binomials obtained in this fashion and by $in_\succ(\mathcal{C})$ the set of their initial monomials. For instance, if $d=5$ then the set $\mathcal{C}$ equals

$$\begin{aligned}\{&\underline{x_{12}x_{34}} - x_{13}x_{24},\ \underline{x_{14}x_{23}} - x_{13}x_{24},\ \underline{x_{12}x_{35}} - x_{13}x_{25},\ \underline{x_{15}x_{23}} - x_{13}x_{25},\\ &\underline{x_{12}x_{45}} - x_{14}x_{25},\ \underline{x_{15}x_{24}} - x_{14}x_{25},\ \underline{x_{13}x_{45}} - x_{14}x_{35},\ \underline{x_{15}x_{34}} - x_{14}x_{35},\\ &\underline{x_{23}x_{45}} - x_{24}x_{35},\ \underline{x_{25}x_{34}} - x_{24}x_{35}\}.\end{aligned} \tag{9.2}$$

Here the initial monomials with respect to $\prec$ are underlined.

Theorem 9.1. *The set $\mathcal{C}$ is the reduced Gröbner basis of $I_{\mathcal{A}_d}$ with respect to $\succ$.*

The reduction relation defined by this Gröbner basis amounts to replacing non-crossing edges by crossing edges. This is illustrated in the following diagram:

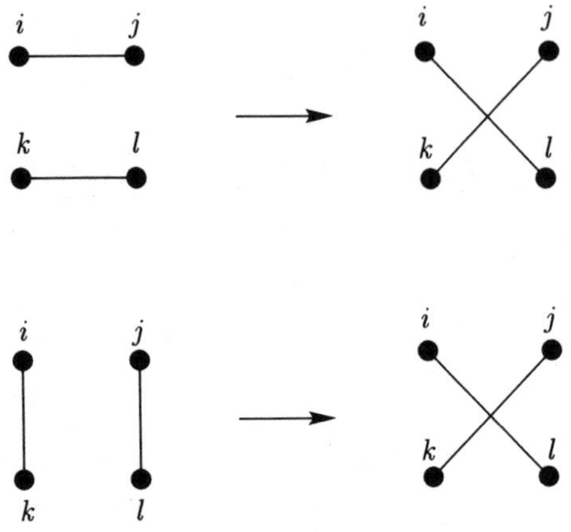

Figure 9-1. Reduction via the Gröbner basis in Theorem 9.1.

Proof of Theorem 9.1: For each binomial $x_{ij}x_{kl} - x_{il}x_{jk}$ in $\mathcal{C}$, the initial term with respect to $\succ$ corresponds to the disjoint edges. This follows from the convex embedding of K_d and the definition of the weights. The integral vectors in the kernel of $\mathcal{A}_d$ are in bijection with even closed walks in the complete graph K_d, and hence so are the binomials of $I_{\mathcal{A}_d}$. More precisely, with an even closed walk $\Gamma = (i_1, i_2, \ldots, i_{2k-1}, i_{2k}, i_1)$ we associate the binomial

$$b_\Gamma \quad := \quad \prod_{l=1}^{k} x_{i_{2l-1}, i_{2l}} - \prod_{l=1}^{k} x_{i_{2l}, i_{2l+1}}$$

Clearly, the walk Γ can be recovered from its binomial b_Γ. By Corollary 4.4, the (infinite) set of binomials associated with all even closed walks in K_d contains every reduced Gröbner basis of $I_{\mathcal{A}_d}$. Therefore in order to prove that $\mathcal{C}$ is a Gröbner basis, it is enough to prove that the initial monomial of any binomial b_Γ is divisible by some monomial $x_{ij}x_{kl}$ where $(i,j),(k,l)$ is a pair of disjoint edges.

Suppose on the contrary there exists a binomial $b_\Gamma \in I_{\mathcal{A}_d}$ that contradicts our assertion. This implies that each pair of edges appearing in the initial monomial of b_Γ intersects. We may assume that b_Γ is a minimal counterexample in the sense that d is minimal and that b_Γ has minimal weight. Here the *weight of a binomial* is the sum of weights of its two terms. The walk Γ is spanning in K_d by the minimality of d. Every edge of Γ gets a label "even" or "odd" according to its position in the walk. If an edge is visited more than once, then it cannot be labeled both odd and even, since otherwise the variable associated with an odd-even edge can be factored out from b_Γ. This would contradict the minimality of weight. Moreover, if $b_\Gamma = \mathbf{x}^\mathbf{u} - \mathbf{x}^\mathbf{v}$ and $in_\succ(b_\Gamma) = \mathbf{x}^\mathbf{u}$, then we can assume that each pair of edges appearing in the trailing monomial $\mathbf{x}^\mathbf{v}$ intersects. Otherwise if $(i,j),(k,l)$ is a non-intersecting pair of edges then we can reduce $\mathbf{x}^\mathbf{v}$ modulo $\mathcal{C}$ to obtain a counterexample of smaller weight.

Let (s,t) be an edge of the walk Γ such that the circular distance between s and t is smallest possible. The edge (s,t) separates the vertices of K_d except s and t into two disjoint sets P and Q where $|P| \geq |Q|$. Let us start

Γ at $(s,t) = (i_1, i_2)$. The walk is then a sequence of vertices and edges $\Gamma = (i_1, (i_1, i_2), i_2, (i_2, i_3), ..., (i_{2k-1}, i_{2k}), i_{2k}, (i_{2k}, i_1))$. Each pair of odd (resp. even) edges intersects. The odd edges are of type (i_{2r-1}, i_{2r}) and the even edges of type (i_{2r}, i_{2r+1}). Since the circular distance of i_1, i_2 is minimal, the vertex i_3 cannot be in Q. Otherwise the edge (i_2, i_3) would have smaller circular distance. We claim that if P contains an odd vertex i_{2r-1}, then it also contains the subsequent odd vertices $i_{2r+1}, i_{2r+3}, \ldots, i_{2k-1}$. The edge (i_1, i_2) is the common boundary of the two regions P and Q. Any odd edge intersects it (at least by having an end in $\{i_1, i_2\}$) and thus i_{2r} is in $Q \cup \{i_1, i_2\}$. Since any even edge must intersect (i_2, i_3), the vertex i_{2r+1} lies in $P \cup \{i_2\}$. To complete the proof of the claim we show that $i_{2r+1} \neq i_2$. The equality $i_{2r+1} = i_2$ would imply either $i_{2r} = i_1$ or $i_{2r} \in Q$. If $i_{2r} = i_1$ then (i_1, i_2) is both odd and even. On the other hand if $i_{2r} \in Q$ then (i_{2r}, i_2) has smaller circular distance than (i_1, i_2). Thus i_{2r+1} belongs to P. The claim is proved by repeating this argument.

Since i_3 was shown to be in P, it follows that all odd vertices except i_1 lie in P and the even vertices lie in $Q \cup \{i_1, i_2\}$. The final vertex i_{2k} is thus in Q. The even edge (i_{2k}, i_1) must be a closed line segment contained in the region Q of the d-gon. Therefore (i_2, i_3) and (i_{2k}, i_1) are two even edges that do not intersect, which is a contradiction. This proves that $\mathcal{C}$ is a Gröbner basis of $I_{\mathcal{A}_d}$ with respect to $\succ$.

By construction, no monomial in an element of $\mathcal{C}$ is divisible by the initial term of an element in $\mathcal{C}$. Hence $\mathcal{C}$ is the reduced Gröbner basis of $I_{\mathcal{A}_d}$ with respect to $\succ$. ■

We remark that the Gröbner basis $\mathcal{C}$ in Theorem 9.1 is actually lexicographic.

Remark 9.2. *The set $\mathcal{C}$ is the reduced Gröbner basis for $I_{\mathcal{A}}$ with respect to the purely lexicographic term order induced by the following variable ordering:*

$$x_{ij} \prec x_{kl} \quad \text{if and only if} \quad i < k \ \text{ or } \ (i = k \text{ and } j > l).$$

Proof: For any ordered quadruple $1 \leq i < j < k < l \leq d$, the intersecting pair of edges is $\{(i,k),(j,l)\}$. We must show that the monomial $x_{ik}x_{jl}$ is smaller than both $x_{ij}x_{kl}$ and $x_{il}x_{jk}$ in the given term order. But this holds since $x_{kl} \succ x_{jk} \succ x_{jl} \succ x_{ij} \succ x_{ik} \succ x_{il}$. ■

We apply Theorem 9.1 to give an explicit triangulation and determine the normalized volume of $\Delta(2,d)$. By Theorem 8.3, the square-free monomial ideal $\langle in_\succ(\mathcal{C})\rangle = in_\succ(I_{\mathcal{A}_d})$ is the Stanley-Reisner ideal of a regular triangulation $\Delta_\succ$ of $\Delta(2,d)$. The simplices in $\Delta_\succ$ are the supports of the standard monomials. All maximal simplices in $\Delta_\succ$ have unit normalized volume (by Corollary 8.9). We observed before that the elements of $in_\succ(\mathcal{C})$, i.e., the minimally non-standard monomials, are supported on pairs of disjoint edges.

Corollary 9.3. *The simplices of $\Delta_\succ$ are the subgraphs of K_d with the property that any pair of edges intersects in the given convex embedding of the graph.*

Here we identify subgraphs of K_d with subpolytopes of $\Delta(2,d)$: a subgraph H is identified with the convex hull of the column vectors of its vertex-edge incidence matrix.

Theorem 9.4. *The maximal simplices in the triangulation $\Delta_\succ$ are spanning subgraphs on d edges with the property that any pair of edges intersects. Every such* *subgraph is connected and contains a unique odd cycle. The number of such subgraphs and hence the normalized volume of the second hypersimplex $\Delta(2,d)$ is $2^{d-1}-d$.*

Lemma 9.5. *A subpolytope σ of $\Delta(2,d)$ is a $(d-1)$-dimensional simplex if and only if the corresponding subgraph H satisfies the following properties:*

(i) H is a spanning subgraph with d edges,

(ii) all cycles in H are odd,

(iii) every component contains at least one odd cycle.

In this case the normalized volume of the simplex σ is $2^{q(H)-1}$ where $q(H)$ is the number of disjoint cycles in H.

Proof of Lemma 9.5: Suppose H supports a $(d-1)$-simplex. Let M_H be the $\{0,1\}$-incidence matrix of H. This matrix is non-singular which implies properties (i) and (ii). Suppose there exists a component C of H with no odd cycles. By property (ii), C is a tree. By induction on the number of edges in the tree one can prove that M_H is singular.

For the converse suppose that (i), (ii) and (iii) hold for H. Then the vertex-edge incidence matrix M_H is square. We shall prove that the absolute value of the determinant of M_H is equal to $2^{q(H)}$. If all vertices of H have degree two, then H is a disjoint union of odd cycles C_i and the matrix M_H (up to permutation of columns) is the direct sum of the matrices M_{C_i}. The determinant of M_H is the product of the determinants of the matrices M_{C_i}. The determinant of the incidence matrix of an odd cycle is 2 or -2. Therefore the absolute value of the determinant of M_H is $2^{q(H)}$. If the set of vertices with degree distinct from two is non-empty, then there is a vertex v of H of degree one. The row associated with v has 1 in some column and 0 elsewhere. Therefore the absolute values of the determinants of M_H and M_{H-v} are equal. Using this repeatedly we can reduce to the first case. The g.c.d. of $\det(M_H)$ where H ranges over all subgraphs of the specified kind is two. Hence the normalized volume of a simplex σ is $2^{q(H)-1}$. ∎

Proof of Theorem 9.4: The characteristics of the subgraphs follow from Corollary 9.3 and Lemma 9.5. Since the normalized volume of a maximal simplex in the triangulation $\Delta_\succ$ is one, we conclude that there is a unique odd cycle in the corresponding subgraph. Recall that the vertices of the graph are the vertices of a regular d-gon numbered in a clockwise manner and the edges are closed line segments joining two vertices. Consider an odd cycle C in K_d with $2k-1$ edges, $k \in \{2,\ldots,\lceil d/2 \rceil\}$. We assume C is drawn such that each pair of edges in C intersect. There are $l = d-(2k-1)$ vertices that are not in C. We need to introduce l new edges in order to obtain a spanning subgraph. Let v be a vertex outside C. Due to the convex embedding of K_d and the requirement that the new edge should intersect all existing edges, there exists a unique vertex w in C such that (v,w) is one of the new edges. Therefore there is exactly one way to complete an odd cycle to a graph with the above properties. There are $\binom{d}{2k-1}$ odd cycles for each $k \in \{2,\ldots,\lceil d/2 \rceil\}$ and hence the total number of such graphs is $\sum_{k=2}^{\lceil d/2 \rceil} \binom{d}{2k-1} = 2^{d-1}-d$. ∎

The graphs appearing as simplices in $\Delta_\succ$ are known as *thrackles* in the combinatorics literature. The standard monomials modulo our Gröbner basis are precisely the *multi-thrackles.* In other words, a monomial $\mathbf{x}^\mathbf{u}$ does not lie in $in_\succ(I_{\mathcal{A}_d})$ if and only if $supp(\mathbf{x}^\mathbf{u})$ is a thrackle. This is equivalent to $\mathbf{x}^\mathbf{u} = x_{i_1j_1}x_{i_2j_2}\cdots x_{i_rj_r}$ where

$$i_1 \le i_2 \le \cdots \le i_r \le j_1 \le j_2 \le \cdots \le j_r, \; i_1 < j_1, \, i_2 < j_2, \, \ldots, i_r < j_r. \qquad (9.3)$$

Note that $\mathbf{x}^\mathbf{u}$ is recovered from $\Phi(\mathbf{x}^\mathbf{u}) = t_{i_1}t_{i_2}\cdots t_{i_r}t_{j_1}t_{j_2}\cdots t_{j_r}$ by simply sorting indices.

The following diagram shows all $11 = 2^{5-1} - 5$ maximal thrackles for $d = 5$. These are the (maximal) supports of the standard monomials modulo the Gröbner basis in (9.2).

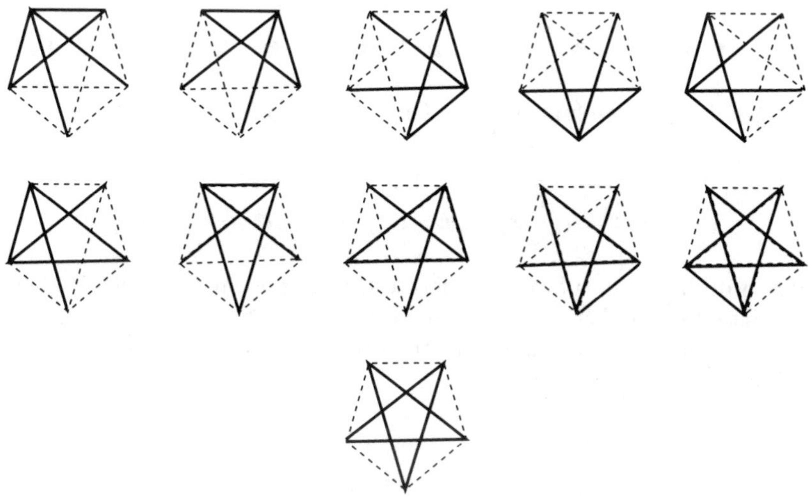

Figure 9-2. Maximal thrackles in a pentagon.

Corollary 9.6. *The Hilbert polynomial of $I_{\mathcal{A}_d}$ equals the Ehrhart polynomial of $\Delta(2,d)$:*

$$H_{\mathcal{A}_d}(r) = card\big(r \cdot \Delta(2,d) \cap \mathbf{Z}^d\big) = \binom{d+2r-1}{d-1} - d \cdot \binom{d+r-2}{d-1}. \qquad (9.4)$$

Proof sketch: We have shown that $\mathcal{A}_d$ has a triangulation into unit simplices. This implies that the Ehrhart polynomial and the Hilbert polynomial are equal. (See also Theorem 13.11 and Proposition 13.15). The right hand expression in (9.4) equals the number of sequences (9.3) and hence the number of standard monomials of degree r. ∎

We next compare the state polytope of $I_{\mathcal{A}_d}$ with the secondary polytope of $\mathcal{A}_d$. Both polytopes lie in $\mathbf{R}^n$ and have dimension $n - d$, where $n = \binom{d}{2}$ is the number of edges in K_d.

Theorem 9.7. *The state polytope of $I_{\mathcal{A}_d}$ and the secondary polytope of $\mathcal{A}_d$ coincide for $d \le 5$ and are distinct for $d \ge 6$.*

Proof: It can be seen by inspection that the matrices $\mathcal{A}_3$, $\mathcal{A}_4$ and $\mathcal{A}_5$ are unimodular. Proposition 8.15 (b) states that in these cases the secondary polytope equals the state polytope.

Before proceeding to $d \geq 6$, let us discuss the cases $d = 4, 5$ in more detail. The hypersimplex $\Delta(2,4)$ is a regular octahedron in $\mathbf{R}^4$. It has three distinct regular triangulations. Therefore the secondary polytope, which is the same as the state polytope, is a triangle in $\mathbf{R}^6$. The three distinct initial ideals of $I_{\mathcal{A}_4}$ are $\langle x_{14}x_{23}, x_{13}x_{24}\rangle$, $\langle x_{13}x_{24}, x_{12}x_{34}\rangle$ and $\langle x_{14}x_{23}, x_{12}x_{34}\rangle$. The hypersimplex $\Delta(2,5)$ has dimension four with 10 vertices and 10 facets (5 tetrahedra and 5 octahedra). Its secondary polytope $\Sigma(\mathcal{A}_5)$ is five-dimensional and has 102 vertices, 255 edges, 240 two-faces, 105 three-faces and 20 facets. Under the natural S_5-action the 102 regular triangulations of $\Delta(2,5)$ fall into three distinct orbits.

Now let $d = 6$ and consider the following subconfiguration of $\mathcal{A}_6$:

$$\mathcal{B} \quad := \quad \{\mathbf{e}_1+\mathbf{e}_2,\, \mathbf{e}_1+\mathbf{e}_6,\, \mathbf{e}_2+\mathbf{e}_3,\, \mathbf{e}_2+\mathbf{e}_6,\, \mathbf{e}_3+\mathbf{e}_4,\, \mathbf{e}_3+\mathbf{e}_5,\, \mathbf{e}_4+\mathbf{e}_5,\, \mathbf{e}_5+\mathbf{e}_6\}. \tag{9.5}$$

Using Algorithm 7.2, we find that the Graver basis of $\mathcal{B}$ equals

$$\begin{aligned} Gr_{\mathcal{B}} \quad = \quad \{ & x_{23}x_{56} - x_{26}x_{35},\ x_{12}x_{56}x_{34} - x_{16}x_{23}x_{45}, \\ & x_{12}x_{56}^2x_{34} - x_{16}x_{26}x_{35}x_{45},\ x_{12}x_{26}x_{35}x_{34} - x_{16}x_{23}^2x_{45}\}. \end{aligned} \tag{9.6}$$

The first two binomials suffice to generate the toric ideal $I_{\mathcal{B}}$, which is therefore a complete intersection. Let ω be the weight vector which assigns the weight 1 to the variables x_{23} and x_{56} and weight 0 to the other six variables. We can read off the initial ideal from (9.6):

$$in_\omega(I_{\mathcal{B}}) \quad = \quad \langle\, x_{23}x_{56},\ x_{12}x_{56}x_{34} - x_{16}x_{23}x_{45},\ x_{12}x_{56}^2x_{34},\ x_{16}x_{23}^2x_{45}\,\rangle. \tag{9.7}$$

This is not a monomial ideal, and we see that (9.7) has precisely two distinct initial ideals $in_{\prec_1}(in_\omega(I_{\mathcal{B}})) = \langle x_{23}x_{56}, x_{12}x_{56}x_{34}, x_{16}x_{23}^2x_{45}\rangle$ and $in_{\prec_2}(in_\omega(I_{\mathcal{B}})) = \langle x_{23}x_{56}, x_{16}x_{23}x_{45}, x_{12}x_{56}^2x_{34}\rangle$. Both have the same radical

$$Rad\big(in_\omega(I_{\mathcal{B}})\big) \quad = \quad \langle\, x_{23}x_{56},\ x_{12}x_{56}x_{34},\ x_{16}x_{23}x_{45}\,\rangle.$$

This shows that the toric ideal $I_{\mathcal{B}}$ has two distinct initial ideals, namely $in_{\prec_{1,\omega}}(I_{\mathcal{B}})$ and $in_{\prec_{2,\omega}}(I_{\mathcal{B}})$ which have the same radical and therefore define the same triangulation $\Delta_{\prec_{1,\omega}} = \Delta_{\prec_{2,\omega}}$ of $\mathcal{B}$. We conclude that the secondary polytope $\Sigma(\mathcal{B})$ is not equal to the state polytope $State(I_{\mathcal{B}})$; see also Exercise (3) below.

We next show the same result for the second hypersimplex, that is, $\Sigma(\mathcal{A}_6) \neq St(I_{\mathcal{A}_6})$. Let $X := \{x_{12}, x_{16}, x_{23}, x_{26}, x_{34}, x_{35}, x_{45}, x_{56}\}$, the set of variables of $I_{\mathcal{B}}$. Let $v \in \mathbf{R}^{\binom{6}{2}}$ be the weight vector that assigns the weight 0 to the eight variables in X and generic positive weights to the other seven variables. This choice of weights has the elimination property:

$$in_v(I_{\mathcal{A}_6}) \,\cap\, k[X] \quad = \quad I_{\mathcal{B}}. \tag{9.8}$$

Let ω be the weight vector which assigns the weight 1 to the variables x_{23} and x_{56} and weight 0 to the other 13 variables. Let us now consider the weight vector $v + \epsilon \cdot \omega$ where ϵ is a very small positive real number. This weight vector is sufficiently generic to define a regular triangulation $\Delta_{v+\epsilon\omega}$ of $\mathcal{A}_6$. Indeed, inside the subpolytope $conv(\mathcal{B})$ this subdivision agrees with the triangulation Δ_ω above, and outside of $conv(\mathcal{B})$ each cell is a simplex, by the genericity of the non-zero coordinates of v.

To show that $\Sigma(\mathcal{A}_6) \neq St(I_{\mathcal{A}_6})$, it suffices to show that the ideal $in_{v+\epsilon\omega}(I_{\mathcal{A}_6})$ is not a monomial ideal. If it were a monomial ideal, then also its elimination ideal

$$\begin{aligned} in_{v+\epsilon\omega}(I_{\mathcal{A}_6}) \cap k[X] &= in_{v+\epsilon\omega}\big(I_{\mathcal{A}_6} \cap k[X]\big) \\ &= in_\omega\big(in_v(I_{\mathcal{A}_6}) \cap k[X]\big) \;=\; in_\omega(I_\mathcal{B}) \end{aligned} \tag{9.9}$$

would be a monomial ideal. This contradicts our result in (9.7), and it hence completes the proof for K_6. In (9.9) the first equation follows from the elimination property of $v+\epsilon\omega$, the second equation follows from Proposition 1.13, and the third equation from (9.8).

Finally, to establish the assertion for $d > 6$, we use the exact same technique, which is to write $I_{\mathcal{A}_6}$ as an elimination ideal of $I_{\mathcal{A}_d}$. This completes the proof of Theorem 9.7. ∎

We next describe the Graver basis of the ideals $I_{\mathcal{A}_d}$ for $d \leq 8$. Recall that the binomials in $I_{\mathcal{A}_d}$ are identified with even walks in the complete graph K_d. The even walks of minimal support are the *circuits.*

Lemma 9.8. *The circuits of $\mathcal{A}_d$ are the following two types of even walks in K_d: even cycles and pairs of disjoint odd cycles joined by a path.*

Proof: This can be derived from the characterization of column bases of $\mathcal{A}_d$ in Lemma 9.5. See Exercise (4) below. ∎

Theorem 9.9. *The circuits form a universal Gröbner basis of $I_{\mathcal{A}_d}$ for $d \leq 7$. The same statement is not true for $d \geq 8$.*

Proof: The proof is by explicit computation using Algorithm 7.2. The elements of the Graver basis are (up to a relabeling using the natural S_d-action) presented in the table below. The numbers give the cardinality of the S_d-orbit of each binomial.

	Types of binomials	$d=5$	$d=6$	$d=7$	$d=8$
(a)	$x_{12}x_{34} - x_{13}x_{24}$	15	45	105	210
(b)	$x_{12}x_{34}x_{35} - x_{23}x_{45}x_{13}$	15	90	315	840
(c)	$x_{12}x_{34}^2x_{56} - x_{13}x_{23}x_{45}x_{46}$	0	90	630	2520
(d)	$x_{12}x_{34}x_{56} - x_{23}x_{45}x_{16}$	0	60	420	1680
(e)	$x_{12}x_{34}x_{56}x_{37} - x_{23}x_{45}x_{67}x_{13}$	0	0	1260	10080
(f)	$x_{12}x_{34}^2x_{57}x_{56} - x_{23}x_{45}^2x_{67}x_{13}$	0	0	630	2520
(g)	$x_{12}x_{34}x_{56}x_{78} - x_{23}x_{45}x_{67}x_{18}$	0	0	0	2520
(h)	$x_{12}x_{34}^2x_{56}^2x_{78} - x_{23}x_{45}^2x_{67}x_{68}x_{13}$	0	0	0	5040
(i)	$x_{12}x_{34}^2x_{56}x_{78} - x_{23}x_{45}x_{67}x_{48}x_{13}$	0	0	0	10080
(j)	$x_{12}x_{34}x_{56}x_{57}x_{38} - x_{23}x_{45}x_{67}x_{58}x_{13}$	0	0	0	2520
(k)	$x_{12}x_{34}x_{56}x_{47}x_{38} - x_{23}x_{45}x_{46}x_{78}x_{13}$	0	0	0	5040
	TOTAL	30	285	3360	45570

The types (a), (d),(g) are even cycles and hence they are circuits. The types (b),(c),(e),(f), (h), (i) are pairs of odd cycles joined by a path and hence they are circuits. Using Proposition 4.11, this proves that for $d \leq 7$ the circuits, the universal Gröbner basis and the Graver basis coincide. For $d = 8$ there are two additional types (j) and (k) of binomials appearing in the Graver basis. To show that they lie in the universal Gröbner basis, we use Theorem 7.8. Indeed, it is easy to show that the binomials (j) and (k) correspond to edges in their respective fibers. If not, then there would exist another lattice point in the same fiber whose support is contained in the 10 appearing variables. But this is impossible as can be seen by inspecting the corresponding graphs with 10 edges and 8 nodes. ∎

The following diagram shows the two graphs, which correspond to the non-circuits (j) and (k) in the universal Gröbner basis of $I_{\mathcal{A}_8}$.

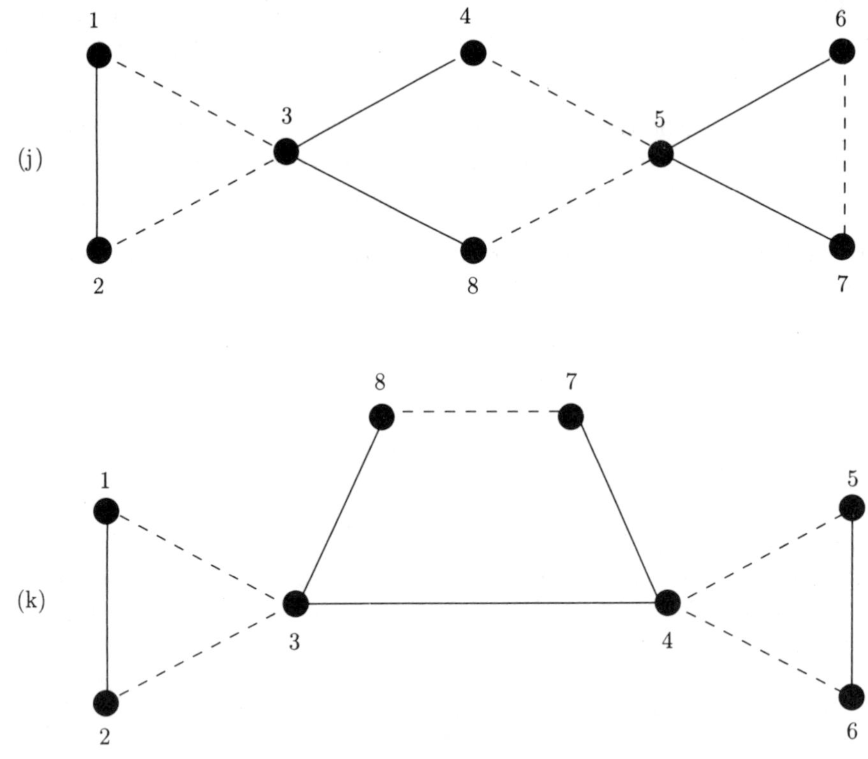

Figure 9-3. The graphs (j) and (k).

The toric ideals $I_{\mathcal{A}_d}$ have the following interpretations in the domains of application we discussed in Chapter 5. The sampling problem for $\mathcal{A}_d$ is the task of generating a random multigraph on d nodes with fixed degree at each node. Here Algorithm 4.2 specializes to the following procedure: start with any legal multigraph and then perform a random walk with respect to a fixed finite set $\mathcal{F}$ of local moves which alters multigraphs while maintaining the vertex degrees. The possible moves are precisely the even walks, that is, binomials in the ideal $I_{\mathcal{A}_d}$. It follows from Theorem 5.3 and Theorem 9.1 that the set of quadratic binomials is sufficient to guarantee connectedness for all fibers.

The task of connecting all fibers becomes much harder if certain edges of K_d are prohibited during the random walk. The algebraic counterpart is to find a generating set for the ideal $I_{\mathcal{B}}$, where $\mathcal{B}$ is any subset of $\mathcal{A}_d$. This problem is solved simultaneously for all subsets of $\mathcal{A}_d$ by finding a universal Gröbner basis for $I_{\mathcal{A}_d}$; see Exercise (2) in Chapter 1. The table in the proof of Theorem 9.9 gives a complete answer for all graphs with $d \leq 8$ nodes. We remark that the subproblem for the class of bipartite graphs is much easier: finding the relevant universal Gröbner basis is the content of Exercise (1) in Chapter 5.

The set of all multigraphs on d nodes with fixed vertex degrees is the set of feasible solutions of a well-known problem of combinatorial optimization, namely, the *perfect f-matching problem.* Let f be a positive integer valued function on the d vertices of K_d such that $f(i)$ specifies the degree of the vertex i. An assignment of a non-negative integer u_{ij} to each edge (i,j) is called a *perfect f-matching* if, for every vertex i, we have $\sum_{j\in\{1,\ldots,n\}\backslash\{i\}} u_{ij} = f(i)$. Suppose in addition to the degree of each vertex we are also given a cost ω_{ij} for each edge (i,j), and the objective is to find a perfect f-matching on d nodes with minimum total cost. This is the *minimum weight perfect f-matching* problem:

$$\begin{aligned} &\text{Minimize } \sum_{i,j} \omega_{ij}\cdot u_{ij} \text{ subject to} \\ &\sum_{j\in\{1,\ldots,n\}\backslash\{i\}} u_{ij} \;=\; f(i),\;\; i=1,\ldots,d,\;\; u_{ij}\in\mathbf{N}. \end{aligned} \tag{9.10}$$

The coefficient matrix of the above integer program is our matrix $\mathcal{A}_d$. We can solve the problem (9.10) using Algorithm 5.6, that is, by computing the reduced Gröbner basis for the toric ideal $I_{\mathcal{A}_d}$ with respect to a term order refining the cost vector $\omega = (\omega_{ij})$.

Exercises:

(1) For $d = 5$, verify Theorem 9.1 by applying Buchberger's criterion to the ten binomials in $\mathcal{C}$. Compute the thrackle triangulation $\Delta_{\prec}$ explicitly in this case.

(2) Compute the number of i-dimensional faces of the second hypersimplex $\Delta(2,d)$.

(3) Let $\mathcal{B}$ be the configuration in (9.5). Prove that $\Sigma(\mathcal{B})$ is a pentagon and that $State(I_{\mathcal{B}})$ is a hexagon.

(4) Prove Lemma 9.8.

(5) Let $F(d)$ denote the largest degree of a binomial in the universal Gröbner basis of $I_{\mathcal{A}_d}$. Show that $d-2 \leq F(d) \leq \binom{d}{2}$.

(6) List all 3-regular labeled multigraphs on six nodes. Use Algorithm 5.7.

(7) Does there exist a quadratic *reverse* lexicographic Gröbner basis for the ideal $I_{\mathcal{A}_d}$?

(8) The r-th *hypersimplex* in $\mathbf{R}^d$ is the convex hull $\Delta(r,d)$ of the configuration

$$\mathcal{A} \quad = \quad \{\mathbf{e}_{i_1} + \mathbf{e}_{i_2} + \cdots + \mathbf{e}_{i_r} \ : \ 1 \leq i_1 < i_2 < \cdots < i_r \leq d\}.$$

What is the normalized volume of $\Delta(r,d)$?

Notes:
The material in this chapter is drawn from (De Loera, Sturmfels, Thomas (1995). The sampling problem of generating multigraphs with a given degree sequence is analyzed in detail in (Sinclair 1993). For more information on the minimum weight perfect f-matching problem – and graph theory in general – consult (Lovasz & Plummer 1986).

CHAPTER 10

$\mathcal{A}$-graded Algebras

What are the graded algebras that have the simplest possible Hilbert function ? This question was raised and partially answered by Arnold (1989) and Korkina, Post & Roelofs (1995). We consider the following multigraded variant of their underlying definition. Let $\mathcal{A} = \{\mathbf{a}_1, \mathbf{a}_2, \ldots, \mathbf{a}_n\}$ be a subset of $\mathbf{N}^d \backslash \{0\}$, $d = dim(\mathcal{A})$, and let $\mathbf{N}\mathcal{A}$ denote the sub-semigroup of $\mathbf{N}^d$ spanned by $\mathcal{A}$. An *$\mathcal{A}$-graded algebra* is a $\mathbf{N}^d$-graded k-algebra $R = \bigoplus_{\mathbf{b}} R_{\mathbf{b}}$ with homogeneous generators $X_1, X_2, \ldots, X_n$ in degrees $\mathbf{a}_1, \mathbf{a}_2, \ldots, \mathbf{a}_n$ such that

$$dim_k(R_{\mathbf{b}}) \quad = \quad \begin{cases} 1 & \text{if } \mathbf{b} \in \mathbf{N}\mathcal{A} \\ 0 & \text{otherwise} \end{cases} \qquad \text{for all } \mathbf{b} \in \mathbf{N}^d. \tag{10.1}$$

Every $\mathcal{A}$-graded algebra R has a natural presentation as a quotient of a polynomial ring:

$$0 \quad \to \quad I \quad \to \quad k[\mathbf{x}] = k[x_1, x_2, \ldots, x_n] \quad \to \quad R \quad \to \quad 0.$$

The presentation ideal $I = ker(x_i \mapsto X_i)$ is called *$\mathcal{A}$-graded* as well. It is easy to see that an $\mathcal{A}$-graded ideal I is generated by polynomials with at most two terms, that is, I is a *binomial ideal.* The paradigm of an $\mathcal{A}$-graded algebra is the semigroup algebra

$$k[\mathbf{N}\mathcal{A}] \quad = \quad k[\mathbf{t}^{\mathbf{a}_1}, \mathbf{t}^{\mathbf{a}_2}, \ldots, \mathbf{t}^{\mathbf{a}_n}] \quad = \quad k[\mathbf{x}]/I_{\mathcal{A}},$$

where $I_{\mathcal{A}}$ is the toric ideal as before.

Two $\mathcal{A}$-graded algebras $R = k[\mathbf{x}]/I$ and $R' = k[\mathbf{x}]/I'$ are considered *isomorphic* if there exists a graded algebra isomorphism of degree 0. This holds if and only if, for the corresponding ideals I and I' in $k[\mathbf{x}]$, there exists $\lambda = (\lambda_1, \ldots, \lambda_n)$ in $(k^*)^n$ such that

$$I' \quad = \quad \lambda \cdot I \; := \; \{ f(\lambda_1 x_1, \ldots, \lambda_n x_n) \; : \; f \in I \}. \tag{10.2}$$

Gröbner basis theory suggests to "pass to the toric limit" in (10.2) as follows:

Remark 10.1. *If I is an $\mathcal{A}$-graded ideal in $k[\mathbf{x}]$ and $\omega \in \mathbf{Z}^n$, then the initial ideal $in_\omega(I)$ is $\mathcal{A}$-graded as well.*

We call an $\mathcal{A}$-graded algebra $R = k[\mathbf{x}]/I$ *coherent* if there exists $\omega \in \mathbf{Z}^n$ such that I is isomorphic to $in_\omega(I_{\mathcal{A}})$. A basic question is whether all $\mathcal{A}$-graded algebras are coherent.

Theorem 10.2. (Arnold 1989; Korkina, Post & Roelofs 1995)
If $d = 1$ and $n = 3$ then every $\mathcal{A}$-graded algebra is coherent.

Theorem 10.2 can be reformulated as follows.

Corollary 10.3. *If $d = 1$ and $n = 3$ then the isomorphism classes of $\mathcal{A}$-graded algebras are in bijection with the faces of the state polytope of the toric ideal $I_{\mathcal{A}}$.*

We will present the proof of Theorem 10.2 in the end of this chapter. First, however, we show that the analogous result does not hold for $d = 1$ and $n \geq 4$. We also present two necessary geometric conditions for coherence, and we construct examples of non-coherent algebras which violate these conditions. The main result in this chapter is Theorem 10.10 which expresses the radicals of $\mathcal{A}$-graded ideals in terms of polyhedral subdivisions.

An $\mathcal{A}$-graded algebra $R = k[\mathbf{x}]/I$ is called *monomial* (or a *mono-AGA*) if its ideal I is generated by monomials. The non-zero monomials of a mono-AGA R are called *standard.* They constitute a k-vector space basis for R.

Theorem 10.4. *Let $d=1, n=4$ and $\mathcal{A}=\{1,3,4,7\}$, and let k be an infinite field.*
(a) There exists a monomial $\mathcal{A}$-graded algebra which is not coherent.
(b) There exists an infinite family of pairwise non-isomorphic $\mathcal{A}$-graded algebras.

Proof: In the polynomial ring $k[x_1, x_2, x_3, x_4]$ we consider the monomial ideal

$$I \quad := \quad \langle\, x_1^3,\, x_1x_2,\, x_2^2,\, x_2x_3,\, x_1x_4,\, x_1^2x_3^2,\, x_1x_3^4,\, x_2x_4^3,\, x_4^4\,\rangle. \qquad (10.3)$$

The quotient algebra $R = k[x_1, x_2, x_3, x_4]/I$ is $\mathcal{A}$-graded. To verify this, one must compute the Hilbert series of I with respect to the grading $deg(x_1) = 1$, $deg(x_2) = 3$, $deg(x_3) = 4$, $deg(x_4) = 7$. This can be done easily using the command `hilb-numer` in the computer algebra system MACAULAY due to Bayer & Stillman (1987b).

We list the standard monomials of low degree:

1	2	3	4	5	6	7	8	9	10	11	12	13	14
x_1	x_1^2	x_2	x_3	x_1x_3	$x_1^2x_3$	x_4	x_3^2	$x_1x_3^2$	x_2x_4	x_3x_4	x_3^3	$x_1x_3^3$	x_4^2
15	16	17	18	19	20	21	22	23	24	25	26	27	28
$x_3^2x_4$	x_3^4	$x_2x_4^2$	$x_3x_4^2$	$x_3^3x_4$	x_3^5	x_4^3	$x_3^2x_4^2$	$x_3^4x_4$	x_3^6	$x_3x_4^3$	$x_3^3x_4^2$	$x_3^5x_4$	x_3^7

The proof is by contradiction. Suppose R is coherent. Then there exists a non-negative vector $\omega = (\omega_1, \omega_2, \omega_3, \omega_4)$ such that $I = in_\omega(I_{\mathcal{A}})$.

(i) In degree 6 we have $x_2^2 \in I$ but $x_1^2x_3 \notin I$. This implies $2\omega_2 > 2\omega_1 + \omega_3$.
(ii) In degree 17 we have $x_1x_3^4 \in I$ but $x_2x_4^2 \notin I$. This implies $\omega_1 + 4\omega_3 > \omega_2 + 2\omega_4$.
(iii) In degree 28 we have $x_4^4 \in I$ but $x_3^7 \notin I$. This implies $4\omega_4 > 7\omega_3$.

Combining these three inequalities we get

$$(2\omega_2) + 2\cdot(\omega_1 + 4\omega_3) + (4\omega_4) \;>\; (2\omega_1 + \omega_3) + 2\cdot(\omega_2 + 2\omega_4) + (7\omega_3). \quad (10.4)$$

The left hand side and the right hand side are both equal to $2\omega_1 + 2\omega_2 + 8\omega_3 + 4\omega_4$. This is a contradiction, and we conclude that R is not coherent. This proves part (a).

To prove part (b) of Theorem 10.4 we consider the following family of ideals:

$$\begin{gathered}\langle\, x_1^2x_3 - c_1x_2^2,\, x_1x_3^4 - c_2x_2x_4^2,\, x_3^7 - c_3x_4^4, \\ x_1^3,\, x_1x_2,\, x_1x_4,\, x_2^3,\, x_2^2x_4,\, x_2x_3,\, x_2x_4^3\,\rangle,\end{gathered} \qquad (10.5)$$

where c_1, c_2, c_3 are indeterminate parameters over k. For every value of c_1, c_2, c_3 in k this is an $\mathcal{A}$-graded ideal in $k[x_1, x_2, x_3, x_4]$. In other words, the given three-dimensional family of ideals is flat over k^3. To see this, we note that the given generators in (10.5) are a Gröbner basis with respect to the lexicographic term order induced from $x_1 > x_2 > x_3 > x_4$. Note also that the three first generators in (10.5) correspond to the three cases (i),(ii) and (iii) above. The ideal I in (10.3) is obtained from (10.5) by a deformation of the form $c_1, c_3 \to \infty, c_2 \to 0$.

Two ideals in this family define isomorphic $\mathcal{A}$-graded algebras if and only if they can be mapped into each other by an element in the torus $(k^*)^4$ (acting naturally on the four variables). This is the case if and only if the invariant c_1c_3/c_2^2 has constant value. We conclude that the ideals in (10.5) define a one-dimensional family of non-isomorphic $\mathcal{A}$-graded algebras. In particular, this family is infinite, since k is infinite. ■

The non-coherent mono-AGA in Theorem 10.4 was found through a systematic search of $\mathcal{A}$-graded monomial algebras. Our point of departure was the following lemma restricting the degrees of minimal generators of an $\mathcal{A}$-graded ideal. Recall from Chapter 4 that a binomial $\mathbf{x^u} - \mathbf{x^v}$ in the toric ideal $I_{\mathcal{A}}$ is called *primitive* if there are no proper monomial factors $\mathbf{x^{u'}}$ of $\mathbf{x^u}$ and $\mathbf{x^{v'}}$ of $\mathbf{x^v}$ such that $\mathbf{x^{u'}} - \mathbf{x^{v'}} \in I_{\mathcal{A}}$. The set of all primitive binomials is the Graver basis $Gr_{\mathcal{A}}$. We say that a vector $\mathbf{b} \in \mathbf{N}\mathcal{A}$ is a *Graver degree* if there exists a binomial $\mathbf{x^u} - \mathbf{x^v}$ in $Gr_{\mathcal{A}}$ having degree $\mathbf{b} = \pi(\mathbf{u}) = \pi(\mathbf{v})$.

Lemma 10.5. *The degree of every minimal generator of an $\mathcal{A}$-graded ideal is a Graver degree.*

Proof: Let I be an $\mathcal{A}$-graded ideal, let f be a homogeneous minimal generator of I, and let $\mathbf{b} := deg(f) \in \mathbf{N}\mathcal{A}$. We must find a primitive binomial $\mathbf{x^u} - \mathbf{x^v}$ of degree $\mathbf{b}$ in $I_{\mathcal{A}}$. By the defining property (10.1), there exists a monomial $\mathbf{x^v}$ of degree $\mathbf{b}$ which is non-zero modulo I. We may assume that f has a minimal number of monomials distinct from $\mathbf{x^v}$. Clearly, this number is at least one, that is, f contains a monomial $\mathbf{x^u}$ distinct from $\mathbf{x^v}$.

We claim that $\mathbf{x^u} - \mathbf{x^v}$ is a primitive binomial in $I_{\mathcal{A}}$. Suppose not, and let $\mathbf{x^{u'}}$ be a proper factor of $\mathbf{x^u}$ and $\mathbf{x^{v'}}$ a proper factor of $\mathbf{x^v}$ such that $\mathbf{x^{u'}}$ and $\mathbf{x^{v'}}$ lie in the same $\mathcal{A}$-graded component of $R = k[\mathbf{x}]/I$. Since $\mathbf{x^{v'}}$ is standard, there exists $c_1 \in k$, such that $\mathbf{x^{u'}} - c_1\mathbf{x^{v'}} \in I$. By the same reasoning, there exists $c_2 \in k$ such that $\mathbf{x^{u-u'}} - c_2\mathbf{x^{v-v'}} \in I$. This implies $\mathbf{x^u} - c_1c_2\mathbf{x^v} \in I$. We may now replace the occurrence of $\mathbf{x^u}$ in f by $c_1c_2\mathbf{x^v}$. This is a contradiction to our minimality assumption, and we are done. ■

Corollary 10.6. *Let $d = 1$ and $\mathcal{A} = \{a_1 < a_2 < \cdots < a_n\} \subset \mathbf{N}$. Then every minimal generator of an $\mathcal{A}$-graded ideal has degree at most $a_{n-1} \cdot a_n$.*

Proof: This follows directly from Lemma 10.5 and Corollary 6.2. ■

If a_{n-1} and a_n are relatively prime, then the bound $a_{n-1} \cdot a_n$ is best possible. To see this note that the binomial $x_n^{a_{n-1}} - x_{n-1}^{a_n}$ appears in the reduced Gröbner basis of $I_{\mathcal{A}}$ with respect to the lexicographic term order induced by $x_1 \succ \cdots \succ x_n$. The initial ideal of $I_{\mathcal{A}}$ for this term order is an $\mathcal{A}$-graded ideal which has a minimal generator of degree $a_{n-1} \cdot a_n$.

The following table comprises a complete catalogue of all non-coherent mono-AGA's for $\mathcal{A} = \{a_1, a_2, a_3, a_4\}$ with $1 \leq a_1 < a_2 < a_3 < a_4 \leq 9$. We write the set $\mathcal{A}$ as a bracket $[a_1a_2a_3a_4]$. The three integers listed immediately after each bracket are:

(i) the number of primitive binomials in $I_{\mathcal{A}}$,
(ii) the total number of all $\mathcal{A}$-graded monomial ideals,
(iii) the number of non-coherent $\mathcal{A}$-graded monomial ideals.

If a quadruple does not appear in this list, then all mono-AGA's are coherent for that $\mathcal{A}$.

[1347]	27	53	2	[1349]	23	38	2	[1456]	26	51	2
[1459]	37	90	10	[1567]	35	79	6	[1568]	27	58	4
[1578]	33	79	2	[1678]	41	112	18	[1689]	32	82	6
[1789]	52	174	42	[2357]	30	75	6	[2358]	31	83	10
[2359]	24	58	8	[2379]	31	82	6	[2567]	30	67	2
[2579]	45	168	42	[2678]	27	53	2	[2689]	23	38	2
[2789]	41	113	10	[3459]	30	63	2	[3479]	31	64	2
[3578]	35	88	2	[3589]	33	81	8	[4569]	32	84	6
[4579]	40	120	6	[5678]	35	90	2	[5789]	40	113	2
[6789]	37	94	6								

Table 10-1. Non-coherent one-dimensional mono-AGA's with $n = 4$ and degrees ≤ 9.

These computational results raise the question whether there exist structural features of coherent AGA's which are not shared by all AGA's. In Propositions 10.8 and 10.9 we shall identify two such features: standard monomials and degrees of minimal generators are subject to certain geometric restrictions in the coherent case.

Observation 10.7. *Every standard monomial* $\mathbf{x}^{\mathbf{u}}$ *of a coherent mono-AGA corresponds to a vertex* $\mathbf{u}$ *of its fiber* $conv(\pi^{-1}(\pi(\mathbf{u})))$.

Proof: Let $\mathbf{x}^{\mathbf{u}}$ be standard in $k[\mathbf{x}]/in_\omega(I_{\mathcal{A}})$. Then $\mathbf{u}$ is the unique point in $\pi^{-1}(\pi(\mathbf{u}))$ at which the linear functional ω attains its minimum. Hence $\mathbf{u}$ is a vertex of its fiber. ∎

The theory of $\mathcal{A}$-graded algebras provides an abstract setting for the study of integer programming problems with respect to a fixed matrix. In this abstract setting we select one lattice point from each fiber. This lattice point is considered "optimal". We say that such a selection rule is an *abstract integer program* if it satisfies the following natural axiom: If $\mathbf{u}$ is the optimal point in its fiber and $\mathbf{v} \leq \mathbf{u}$ componentwise, then $\mathbf{v}$ is the optimal point in its fiber. An abstract integer program is *coherent* if it is induced by a linear functional ω, that is, if it truly corresponds to a family of integer programs as in Chapter 5.

This notion of an abstract integer program is identical to the notion of a monomial $\mathcal{A}$-graded algebra, as defined above. Indeed, a *mono-AGA* can be thought of as a rule which selects one lattice point (called *standard monomial*) from each fiber, subject to the one axiom that the set of standard monomials is closed under divisibility. Clearly, an abstract integer program is coherent if and only if the corresponding *mono-AGA* is coherent.

Does there exist a (non-coherent) mono-AGA which has a standard monomial that is not a vertex of its own fiber ? The answer was found to be "no" for all 218 non-coherent mono-AGA's listed in Table 10-1. We do not know the answer for $d = 1$ and $n = 4$ in general. For $d = 1$ and $n = 5$ we can show that the answer is "yes".

Proposition 10.8. *Let $d = 1, n = 5$ and $\mathcal{A} = \{3,4,5,13,14\}$. There exists a monomial $\mathcal{A}$-graded algebra which has a standard monomial $\mathbf{x}^{\mathbf{u}}$ such that $\mathbf{u}$ is not a vertex of its fiber $conv(\pi^{-1}(\pi(\mathbf{u})))$.*

Proof. In the polynomial ring $k[x_1, x_2, x_3, x_4, x_5]$ we consider the ideal

$$I \quad = \quad \langle\, x_1^3,\, x_2^2,\, x_3^2,\, x_1x_5,\, x_2x_5,\, x_3x_5,\, x_5^2\,\rangle.$$

This ideal is $\mathcal{A}$-graded. Indeed, an easy MAPLE or MACAULAY computation shows that $R = k[\mathbf{x}]/I$ has the correct Hilbert series, namely,

$$\frac{1}{1-t} - t - t^2 \quad = \quad \sum_{m \in \mathbf{N}\mathcal{A}} t^m.$$

The monomial $x_1^2x_2x_3$ does not lie in I: it is a standard monomial of degree 15. The corresponding fiber consists of precisely four monomials: $\pi^{-1}(15) = \{x_1^5, x_1x_2^3, x_1^2x_2x_3, x_3^3\}$. The convex hull of $\pi^{-1}(15)$ is the triangle

$$conv\,\{(5,0,0,0,0),(1,3,0,0,0),(0,0,3,0,0)\}.$$

The point $(2,1,1,0,0)$ lies in the relative interior of this triangle. ∎

We recall that an element $\mathbf{b}$ of $\mathbf{N}\mathcal{A}$ is a *Gröbner degree* if a binomial of degree $\mathbf{b}$ appears in some reduced Gröbner basis of $I_{\mathcal{A}}$. These degrees were characterized geometrically in terms of edges of fibers in Corollary 7.12. Clearly, the degree of every minimal generator of a coherent $\mathcal{A}$-graded ideal is a Gröbner degree. The following example shows that this does not hold in general. Note that the number $n = 145$ is certainly not best possible.

Proposition 10.9. *For $d = 1$ and $n = 145$ there exists an $\mathcal{A}$-graded ideal I which has a minimal generator whose degree is not a Gröbner degree.*

Proof: Let $\mathcal{A}' = \{15, 20, 23, 24, 107, 109\}$. Let S' be a polynomial ring in six variables $x_{15}, x_{20}, x_{23}, x_{24}, x_{107}, x_{109}$. We grade S' by setting $deg(x_i) = i$. Let M' be the ideal generated by the six variables, and let $M'_{\geq 139}$ be the ideal generated by all monomials of degree ≥ 139 in S'. Let I be the binomial ideal generated by

$$x_{15}^4,\ x_{20}^2x_{23},\ x_{15}^3x_{23},\ x_{15}^3x_{24},\ x_{24}x_{23}^2,\ x_{15}x_{23}^3,\ x_{20}x_{24}^3,\ x_{15}x_{20}^4,\ x_{15}^2x_{20}x_{23}^2,$$
$$x_{24}^5,\ x_{15}x_{93},\ x_{15}x_{109},\ x_{24}x_{109},\ x_{15}^2x_{107},\ \text{and}\ x_{15}^2x_{20}^3x_{24}^2 - x_{23}^6.$$

The binomial $x_{15}^2x_{20}^3x_{24}^2 - x_{23}^6$ has degree 138. This degree is Graver but not Gröbner, by Example 7.11 and Proposition 4.13 (c),(d). The ideal I is constructed to have the following property: the Artinian ring $S'/(I + M'_{\geq 139})$ is $\mathcal{A}$-graded up to degree 138. In other words, its Hilbert series equals $\sum\{\, t^b \,:\, b \in \mathbf{N}\mathcal{A} \text{ and } b \leq 138\}$.

Let $\mathcal{A}'' = \{139, 140, \ldots, 277\}$ and introduce the corresponding polynomial ring $S'' = k[x_{139}, x_{140}, \ldots, x_{277}]$. We write M'' for the ideal generated by all 139 variables in S'', and we let J'' be any $\mathcal{A}''$-graded ideal in S''.

Finally, we set $\mathcal{A} := \mathcal{A}' \cup \mathcal{A}''$, and we introduce the corresponding 145-variate polynomial ring $S := S' \otimes_k S''$. In this ring we form the ideal

$$J \quad := \quad \langle M' \cdot M'' \rangle \,+\, \langle I + M'_{\geq 139} \rangle \,+\, \langle J'' \rangle.$$

By construction, the ideal J is $\mathcal{A}$-graded in S. It has the primitive binomial $x_{15}^2 x_{20}^3 x_{24}^2 - x_{23}^6$ among its minimal generators. However, its degree 138 is not a Gröbner degree for $\mathcal{A}$. This completes the proof. ∎

We now present a polyhedral construction for the radical of an arbitrary $\mathcal{A}$-graded ideal. If σ is any subset of $\mathcal{A}$, then we identify its toric ideal I_σ with the prime ideal $I_\sigma + \langle x_i : \mathbf{a}_i \notin \sigma \rangle$ in $k[\mathbf{x}]$. A *polyhedral subdivision* of $\mathcal{A}$ is a collection Δ of subsets of $\mathcal{A}$ such that $\{\, pos(\sigma) \,:\, \sigma \in \Delta\}$ is a polyhedral fan with support equal to the cone $pos(\mathcal{A})$. In Chapter 8 we considered the special case where each $\sigma \in \Delta$ has cardinality $d = dim(\mathcal{A})$, in which case Δ is a triangulation of $\mathcal{A}$. A basic construction due to Stanley (1987) associates to any integral polyhedral complex a radical binomial ideal. If Δ is a polyhedral subdivision of $\mathcal{A}$, then its *Stanley ideal* is $I_\Delta := \bigcap_{\sigma \in \Delta} I_\sigma$. We remark that $R = k[\mathbf{x}]/I_\Delta$ is also graded by the semigroup $\mathbf{N}\mathcal{A}$, but it is generally <u>not</u> $\mathcal{A}$-graded. The reason is that for some $\mathbf{b} \in \mathbf{N}\mathcal{A}$, the graded component $R_{\mathbf{b}}$ may be zero. Finally, we say that two arbitrary ideals I and I' in $k[\mathbf{x}]$ are *torus isomorphic* if there exists $\lambda \in (k^*)^n$ such that (10.2) holds.

Theorem 10.10. *If I is any $\mathcal{A}$-graded ideal, then there exists a polyhedral subdivision Δ of $\mathcal{A}$ such that $Rad(I) = \bigcap_{\sigma \in \Delta} J_\sigma$ where each component J_σ is a prime ideal torus isomorphic to I_σ.*

Before presenting the proof of this theorem let us make two remarks. First, as a special case of Theorem 10.10, we can recover Theorem 8.3 and Corollary 8.4. Namely, this is the special case where the given ideal I is a coherent mono-AGA. Our second remark is to explain the mysterious appearance of the ideals J_σ. What is the point of replacing I_σ by a torus isomorphic ideal J_σ, for each maximal cell σ of Δ ? The answer is that the Stanley ideal $I_\Delta = \bigcap_{\sigma \in \Delta} I_\sigma$ itself is generally <u>not</u> torus isomorphic to the radical of I.

Example 10.11. *An $\mathcal{A}$-graded ideal I with $Rad(I)$ not torus isomorphic to I_Δ.* Let $d = 3, n = 6$ and $\mathcal{A} = \{(4,0,0),(0,4,0),(0,0,4),(2,1,1),(1,2,1),(1,1,2)\}$ as in Example 8.2. For every choice of constants $c_1, c_2, c_3 \in k^*$, the following ideal is $\mathcal{A}$-graded:

$$\begin{aligned} I_{c_1, c_2, c_3} \quad = \quad & \langle x_1x_2x_3,\ x_1x_5x_6,\ x_2x_4x_6,\ x_3x_4x_5,\ x_1x_2x_6^2,\ x_1x_3x_5^2,\ x_2x_3x_4^2, \\ & x_1x_5^4 - c_1x_2x_4^4,\ x_2x_6^4 - c_2x_3x_5^4,\ x_3x_4^4 - c_3x_1x_6^4 \rangle. \end{aligned}$$

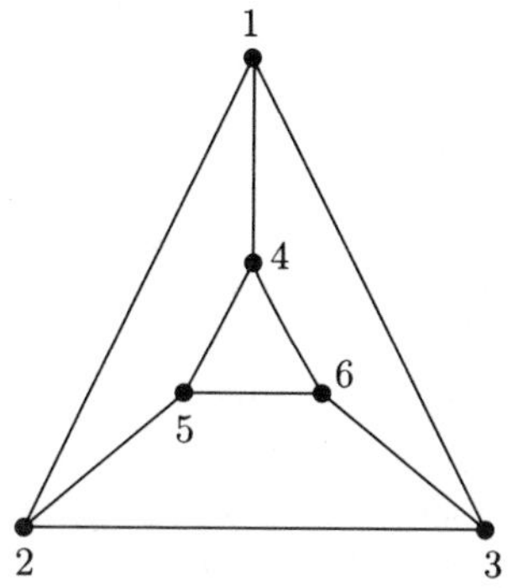

Figure 10-1. The subdivision underlying Example 10.11.

The underlying subdivision Δ of $\mathcal{A}$ consists of three quadrangular cones and one triangular cone. This can be seen from the prime decomposition

$$\begin{aligned} Rad(I_{c_1,c_2,c_3}) &= \langle x_1, x_2, x_3 \rangle \cap \langle x_1 x_5^4 - c_1 x_2 x_4^4, x_3, x_6 \rangle \\ &\cap \langle x_2 x_6^4 - c_2 x_3 x_5^4, x_1, x_4 \rangle \cap \langle x_3 x_4^4 - c_3 x_1 x_6^4, x_2, x_5 \rangle. \end{aligned}$$

From this decomposition we see that $Rad(I_{c_1,c_2,c_3})$ is torus isomorphic to the Stanley ideal I_Δ if and only if the invariant $c_1 c_2 c_3$ attains the value 1. The reader will not fail to note a certain analogy between this example and the three-dimensional family in (10.5).

The deeper reason for the phenomenon in Example 10.11 is the existence of moduli (infinite families) of isomorphism classes of $\mathcal{A}$-graded algebras. These moduli stem from "extraneous components" in the parameter space of AGA's. The structure of this parameter space will be studied in a future publication.

For the proof of Theorem 10.10 we need the following lemma.

Lemma 10.12. *Let $I \subset k[\mathbf{x}]$ be an $\mathcal{A}$-graded ideal which contains no monomials. Then I is torus isomorphic to the toric ideal $I_\mathcal{A}$.*

Proof: For every vector $\mathbf{u} \in ker(\pi)$ there exists a unique non-zero constant $\gamma_\mathbf{u} \in k^*$ such that $\mathbf{x}^{\mathbf{u}^+} - \gamma_\mathbf{u} \cdot \mathbf{x}^{\mathbf{u}^-}$ lies in the ideal I. The fact that I contains no monomials implies that $\gamma_{-\mathbf{u}} = \gamma_\mathbf{u}^{-1}$ and $\gamma_{\mathbf{u}+\mathbf{v}} = \gamma_\mathbf{u} \cdot \gamma_\mathbf{v}$ for all $\mathbf{u}, \mathbf{v} \in ker(\pi)$. In other words, the map $ker(\pi) \to k^*, \mathbf{u} \mapsto \gamma_\mathbf{u}$ is a homomorphism of abelian groups. The abelian group $ker(\pi)$ is a direct summand of $\mathbf{Z}^n$. Therefore there exists a homomorphism $\gamma : \mathbf{Z}^n \to k^*$ such that $\gamma(\mathbf{u}) = \gamma_\mathbf{u}$ for all $\mathbf{u} \in ker(\pi)$. Now define $\lambda = (\lambda_1, \ldots, \lambda_n) \in (k^*)^n$ by $\lambda_i := \gamma(\mathbf{e}_i)$, the image of the i-th unit vector. Then $\lambda \cdot I = I_\mathcal{A}$ as desired. ∎

Proof of Theorem 10.10: The polynomial ring $S = k[x_1, \ldots, x_n]$ is graded by the semigroup $\mathbf{N}\mathcal{A}$ via $deg(x_i) = \mathbf{a}_i$. For any $\mathbf{b} \in \mathbf{N}\mathcal{A}$ we define the subalgebra $S_{(\mathbf{b})} := \bigoplus_{m=0}^\infty S_{m\mathbf{b}}$. This algebra is generated by a finite set of monomials. Inside it we consider the binomial ideal $I_{(\mathbf{b})} := I \cap S_{(\mathbf{b})}$. The corresponding subalgebra $R_{(\mathbf{b})} := S_{(\mathbf{b})}/I_{(\mathbf{b})}$ of our given $\mathcal{A}$-graded algebra $R = S/I$ is a finitely generated k-algebra of Krull dimension 1. It is not possible that all elements in such an algebra are nilpotent. We conclude that there exists a monomial $\mathbf{x^u}$ in $S_{(\mathbf{b})}$ which is not nilpotent modulo $I_{(\mathbf{b})}$.

Let $\mathbf{x^u}$ and $\mathbf{x^v}$ be two such non-nilpotent monomials in $R_{(\mathbf{b})}$. We claim that their product $\mathbf{x^u x^v} \in R_{(\mathbf{b})}$ is not nilpotent either. To see this, we choose integers m_1 and m_2 such that $x^{m_1 \mathbf{u}}$ and $x^{m_2 \mathbf{v}}$ have the same degree. There exists a non-zero constant $c \in k^*$ such that $\mathbf{x}^{m_1\mathbf{u}} = c \cdot \mathbf{x}^{m_2 \mathbf{v}}$ in $R_{(\mathbf{b})}$. This implies $(\mathbf{x}^{m_1\mathbf{u}}\mathbf{x}^{m_2\mathbf{v}})^m = c^m(\mathbf{x^v})^{2mm_2} = c^{-m}(\mathbf{x^u})^{2mm_1} \neq 0$ in R for all $m > 0$, and consequently $(\mathbf{x^u x^v})^m \neq 0$ in $R_{(\mathbf{b})}$ for all $m > 0$. We have shown that the set of non-nilpotent monomials in $R_{(\mathbf{b})}$ is multiplicatively closed.

This multiplicativity property allows us to synthesize the polyhedral subdivision Δ. The *support* of a monomial is defined as $supp(\mathbf{x^u}) := \{\mathbf{a}_i \in \mathcal{A} : u_i \neq 0\}$. Clearly, we have $supp(\mathbf{x^u x^v}) = supp(\mathbf{x^u}) \cup supp(\mathbf{x^v})$. This implies that the set of supports of non-nilpotent monomials in $R_{(\mathbf{b})}$ has a unique maximal element. This subset of $\mathcal{A}$ is denoted $cell(\mathbf{b})$. We define Δ to be the collection of all subsets $cell(\mathbf{b})$ as $\mathbf{b}$ ranges over $\mathbf{N}\mathcal{A}$.

We shall prove that Δ is indeed a polyhedral subdivision of $\mathcal{A}$. Let τ be any face of $\sigma = cell(\mathbf{b})$ (possibly $\tau = \sigma$), and let $\mathbf{b}'$ be any lattice point in the relative interior of $pos(\tau)$. It suffices to show that $cell(\mathbf{b}') = \tau$. By the property of being a face, τ is the unique maximal subset of σ which is the support of any monomial $\mathbf{x}^{\mathbf{u}'}$ in $S_{(\mathbf{b}')}$. Such a monomial $\mathbf{x}^{\mathbf{u}'}$ is not nilpotent modulo I, since there exists a monomial $\mathbf{x^u}$ in $S_{(\mathbf{b})}$ whose support equals $\sigma \supset \tau = supp(\mathbf{x}^{\mathbf{u}'})$. Suppose there exists a non-nilpotent monomial $\mathbf{x}^{\mathbf{u}''}$ in $R_{(\mathbf{b}')}$ whose support $\rho := supp(\mathbf{x}^{\mathbf{u}''})$ properly contains τ. Then $\rho \backslash \sigma = \rho \backslash \tau \neq \emptyset$. Choose integers m_1, m_2 and a non-zero constant $c \in k^*$ such that $\mathbf{x}^{m_1\mathbf{u}'} - c \cdot \mathbf{x}^{m_2\mathbf{u}''} \in I$. Let the degree of this binomial be $m_3 \cdot \mathbf{b}'$. Choose an integer $m_4 \gg 0$ such that $m_4 \cdot \mathbf{b} - \mathbf{b}'$ lies in the relative interior of $pos(\sigma)$, and let $\mathbf{x^w}$ be a monomial having degree $m_4 \cdot \mathbf{b} - \mathbf{b}'$ and support σ. We conclude that $\mathbf{x}^{m_3\mathbf{w}+m_1\mathbf{u}'} - c \cdot \mathbf{x}^{m_3\mathbf{w}+m_2\mathbf{u}''}$ lies in the degree $m_3 m_4 \cdot \mathbf{b}$ component of the ideal I. The first monomial $\mathbf{x}^{m_3\mathbf{w}+m_1\mathbf{u}'}$ is not nilpotent modulo I since it has support σ. The second monomial $\mathbf{x}^{m_3\mathbf{w}+m_2\mathbf{u}''}$ is nilpotent modulo I since its support $\sigma \cup \rho$ strictly contains σ. This is a contradiction, and we conclude that $cell(\mathbf{b}') = \tau$. This completes the proof that Δ is a polyhedral subdivision of $\mathcal{A}$.

We next compute the radical of I. Let σ be a maximal cell in Δ. By construction, the elimination ideal $I \cap k[x_i : i \in \sigma]$ is a σ-graded ideal which contains no monomials. Lemma 10.12 implies that its natural embedding into $k[\mathbf{x}]$,

$$J_\sigma \quad := \quad (I \cap k[x_i : i \in \sigma]) \; + \; \langle x_j \; : \; j \notin \sigma \rangle,$$

is torus isomorphic to the toric prime I_σ. We claim that $Rad(I) = \bigcap_{\sigma \in \Delta} J_\sigma$.

We first show the inclusion $I \subseteq \bigcap_{\sigma\in\Delta} J_\sigma$. (This automatically implies $Rad(I) \subseteq \bigcap_{\sigma\in\Delta} J_\sigma$ because the right hand side is radical.) If $\mathbf{x^u}$ is any monomial not contained in $\bigcap_{\sigma\in\Delta}\langle x_j : j \notin \sigma\rangle$, then $supp(\mathbf{x^u}) \subseteq \sigma$ for some $\sigma \in \Delta$, and hence $\mathbf{x^u}$ is not nilpotent modulo I. This shows that all monomials which are nilpotent modulo I lie in $\bigcap_{\sigma\in\Delta} J_\sigma$. Consider any binomial $f := \mathbf{x^u} - c \cdot \mathbf{x^v}$ in I with both terms not nilpotent modulo I. Let $\mathbf{b} = deg(\mathbf{x^u}) = deg(\mathbf{x^v})$. Fix $\sigma \in \Delta$. If $cell(\mathbf{b})$ is a face of σ, then $f \in I_{(\mathbf{b})} \cap k[x_i : i \in \sigma] \subset J_\sigma$. If $cell(\mathbf{b})$ is not a face of σ, then the supports of $\mathbf{x^u}$ and $\mathbf{x^v}$ are not subsets of σ. Therefore both $\mathbf{x^u}$ and $\mathbf{x^v}$ lie in $\langle x_j : j \notin \sigma\rangle$, and hence $f \in J_\sigma$.

For the reverse inclusion $\bigcap_{\sigma\in\Delta} J_\sigma \subseteq Rad(I)$ we use the Nullstellensatz: it suffices to prove that the variety $\mathcal{V}(I)$ is contained in $\cup_\sigma \mathcal{V}(J_\sigma)$. Let $\mathbf{u} \in \bar{k}^n$ be any zero of I, where $\bar{k}$ is the algebraic closure of k. Abbreviate $\rho := supp(\mathbf{u})$. Consider the monomial $\prod_{i\in\rho} x_i$ and let $\mathbf{b}$ be its degree. Let σ be any maximal cell of Δ which has $cell(\mathbf{b})$ as a face. By construction, no power of $\prod_{i\in\rho} x_i$ vanishes at $\mathbf{u}$. Hence the monomial $\prod_{i\in\rho} x_i$ is not nilpotent modulo I, and its support ρ is a subset of $cell(\mathbf{b}) \subseteq \sigma$. In other words, $\mathbf{u}$ is a zero of the ideal $\langle x_j : j \notin \sigma \rangle$. Therefore $\mathbf{u}$ is a zero of J_σ. This completes the proof. ∎

The natural question arises whether the converse to Theorem 10.10 holds, i.e., whether every polyhedral subdivision of a set $\mathcal{A}$ appears as the reduced scheme of some $\mathcal{A}$-graded algebra. This question was answered to the negative by Irena Peeva.

Theorem 10.13. (Peeva, personal communication)
Let $d = 4$ and $n = 7$. There exists a (non-regular) triangulation Δ of the set $\mathcal{A} = \{(1,0,0,1), (0,1,0,1), (0,0,1,1), (2,0,0,1), (0,2,0,1), (0,0,2,1), (1,1,1,1)\}$ such that the Stanley ideal I_Δ is not the radical of any $\mathcal{A}$-graded ideal.

Proof: The polytope $Q = conv(\mathcal{A})$ is a capped triangular prism. Its vertex set equals $\mathcal{A}$. In (Lee 1991) we find the following non-regular triangulation:

$$\Delta \quad = \quad \{1237,\, 1257,\, 1347,\, 1457,\, 2367,\, 2567,\, 3467\}.$$

The Stanley ideal of Δ equals $I_\Delta = \langle x_1x_6,\, x_2x_4,\, x_3x_5,\, x_4x_5x_6\rangle$. Suppose there exists an $\mathcal{A}$-graded ideal I such that $Rad(I) = I_\Delta$. We may assume that I is a monomial ideal; if not, replace I by any of its initial ideals. This is legal because $Rad(in_\prec(I)) = Rad(in_\prec(Rad(I))) = Rad(in_\prec(I_\Delta)) = I_\Delta$. Consider the following three relations:

$$\underline{x_1^2x_6} - x_3^2x_4,\quad \underline{x_2^2x_4} - x_1^2x_5,\quad \underline{x_3^2x_5} - x_2^2x_6 \quad \in \quad I_{\mathcal{A}}.$$

The non-underlined terms do not lie in I_Δ and hence they do not lie in I. Since I is a monomial ideal, we conclude that the three underlined terms lie in I. The fiber over $\mathbf{b} = (2,2,2,5)$ consists of precisely three monomials: $\underline{x_1^2x_6}x_2^2$, $\underline{x_2^2x_4}x_3^2$, and $\underline{x_3^2x_5}x_1^2$. All three of them lie in I. This is a contradiction to $dim_k(k[\mathbf{x}]/I)_\mathbf{b} = 1$. ∎

However, the converse to Theorem 10.10 does hold in one important special case, namely, when $\mathcal{A}$ is unimodular and Δ is a triangulation.

Lemma 10.14. *If $\mathcal{A}$ is unimodular, then the map $\Delta \mapsto I_\Delta$ defines a bijection between the set of triangulations of $\mathcal{A}$ and the set of $\mathcal{A}$-graded monomial ideals. Moreover, the triangulation Δ is regular if and only if the corresponding $\mathcal{A}$-graded ideal I_Δ is coherent.*

Proof: Let Δ be any triangulation of $\mathcal{A}$ and $I_\Delta = \cap_{\sigma\in\Delta}\langle x_j : j \notin \sigma\rangle$ its Stanley ideal in $k[\mathbf{x}]$. We will show that I_Δ is $\mathcal{A}$-graded. Let $\mathbf{b} \in \mathbf{N}\mathcal{A}$. There exists a unique cone σ in Δ which contains $\mathbf{b}$. We can write $\mathbf{b} = \sum_{i\in\sigma} \lambda_i \mathbf{a}_i$, where λ_i are positive rationals. Since $\mathcal{A}$ is unimodular, the coefficients λ_i are in fact integers. We conclude that $\prod_{i\in\sigma} x_i^{\lambda_i}$ is the unique monomial of degree $\mathbf{b}$ which does not lie in I_Δ. This proves that I_Δ is $\mathcal{A}$-graded.

We have shown that $\Delta \mapsto I_\Delta$ is an injective map from the set of triangulations of $\mathcal{A}$ into the set of $\mathcal{A}$-graded monomial ideals. To see that it is surjective, let I

be any $\mathcal{A}$-graded monomial ideal. By Theorem 10.10, there exists a triangulation Δ of $\mathcal{A}$ such that $Rad(I) = I_\Delta$. By the previous paragraph, the ideal I_Δ is $\mathcal{A}$-graded. Hence $I \subseteq I_\Delta$ is an inclusion of two ideals with the same Hilbert function. This implies $I = I_\Delta$, as desired. The second assertion in Lemma 10.14 follows immediately from Theorem 8.3. ∎

An application of Lemma 10.14 is the study of triangulations of the product of two simplices. Here $\mathcal{A}$ is the configuration in Example 5.1 and the toric ideal $I_\mathcal{A}$ is generated by the 2×2-minors of an $r \times s$-matrix of indeterminates. This configuration $\mathcal{A}$ is known to be unimodular, by Exercise 9 in Chapter 8.

Theorem 10.15. *The product of two simplices $\Delta_{r-1} \times \Delta_{s-1}$ possesses a non-regular triangulation if and only if $(r-2)(s-2) \geq 4$.*

Proof: If $s = 2$ then all triangulations are regular by (Gel'fand, Kapranov & Zelevinsky 1994; Chapter 7, Proposition 3.10 (b)). We may therefore assume $r \geq s \geq 3$. We shall make use of the fact that if there is a non-regular triangulation for (r,s) then there are non-regular triangulations for $(r+1,s)$ and for $(r,s+1)$. (Reason: every triangulation of one face of a polytope can be extended to a triangulation of the whole polytope.)

For the only-if direction, let $(r-2)(s-2) < 4$. This is equivalent to $(r,s) \in \{(3,3),(4,3),(5,3)\}$. It suffices to show that all triangulations are regular when $(r,s) = (5,3)$. Using a modification of the MAPLE program which generated Table 10-1, we computed all mono-AGA's for this case (See Exercise (4)). Modulo the natural $S_5 \times S_3$-symmetry (by permuting rows and columns of the 5×3-matrix (x_{ij})), we found precisely 530 $\mathcal{A}$-graded monomial ideals. Lemma 10.14 implies that there are 530 symmetry classes of triangulations of $\Delta_4 \times \Delta_2$. All of them were shown to be regular by De Loera (1995a).

For the proof of the if-direction it suffices to consider the cases $(r,s) = (4,4)$ and $(r,s) = (6,3)$. The first case is taken care of by De Loera's non-regular triangulation of the product of two tetrahedra in Example 8.13. What remains is to construct a non-regular triangulation of $\Delta_5 \times \Delta_2$. The $\mathcal{A}$-grading of the toric ideal $I_\mathcal{A} = \langle x_{ij}x_{kl} - x_{il}x_{kj} : 1 \leq i < j \leq 6, 1 \leq k < l \leq 3 \rangle$ is given by $deg(x_{ij}) = y_i z_j$. We fix the weight matrix

$$\omega \quad = \quad \begin{pmatrix} 5 & -38 & -13 \\ 7 & -6 & 19 \\ 13 & -30 & -2 \\ 17 & 4 & 2 \\ 11 & -5 & 23 \\ 19 & 3 & 1 \end{pmatrix}$$

The initial ideal of $I_\mathcal{A}$ with respect to these weights equals $in_\omega(I_\mathcal{A}) =$

$$\langle\ \underline{x_{11}x_{32}} - x_{31}x_{12},\quad \underline{x_{61}x_{13}} - x_{11}x_{63},\quad \underline{x_{12}x_{23}} - x_{13}x_{22},$$
$$\underline{x_{21}x_{53}} - x_{51}x_{23},\quad \underline{x_{41}x_{22}} - x_{21}x_{42},\quad \underline{x_{31}x_{43}} - x_{41}x_{33},$$
$$\underline{x_{33}x_{52}} - x_{32}x_{53},\quad \underline{x_{42}x_{63}} - x_{43}x_{62},\quad \underline{x_{51}x_{62}} - x_{61}x_{52},$$
$$x_{11}x_{22},\ x_{11}x_{23},\ x_{11}x_{33},\ x_{11}x_{42},\ x_{11}x_{43},\ x_{11}x_{52},\ x_{11}x_{53},\ x_{11}x_{62},\ x_{12}x_{33},$$
$$x_{12}x_{53},\ x_{13}x_{42},\ x_{13}x_{62},\ x_{22}x_{33},\ x_{22}x_{53},\ x_{23}x_{42},\ x_{23}x_{62},\ x_{31}x_{22},\ x_{31}x_{23},$$
$$x_{31}x_{42},\ x_{31}x_{52},\ x_{31}x_{53},\ x_{31}x_{62},\ x_{33}x_{42},\ x_{33}x_{62},\ x_{41}x_{23},\ x_{41}x_{53},\ x_{42}x_{53},\ x_{51}x_{22},$$
$$x_{51}x_{42},\ x_{53}x_{62},\ x_{61}x_{22},\ x_{61}x_{23},\ x_{61}x_{33},\ x_{61}x_{42},\ x_{61}x_{43},\ x_{61}x_{53},\ x_{13}x_{41}x_{52}\,\rangle.$$

Clearly, this (non-monomial) ideal is $\mathcal{A}$-graded. We now define J to be the ideal generated by the nine underlined monomials together with the 37 other monomials (36 quadrics and one cubic). The passage from $in_\omega(I_\mathcal{A})$ to J does not alter the Hilbert function. (This needs to be checked by computer.) Therefore J is an $\mathcal{A}$-graded monomial ideal. By direct inspection we see that J is not coherent: the above nine binomials have the property that the product of the underlined terms is equal to the product of the non-underlined trailing terms. Thus there is no term order which selects the underlined terms as initial terms. By Lemma 10.14, the ideal J defines a non-regular triangulation of $\Delta_5 \times \Delta_2$. ∎

We remark that the previous example was derived from the *Pappus configuration*:

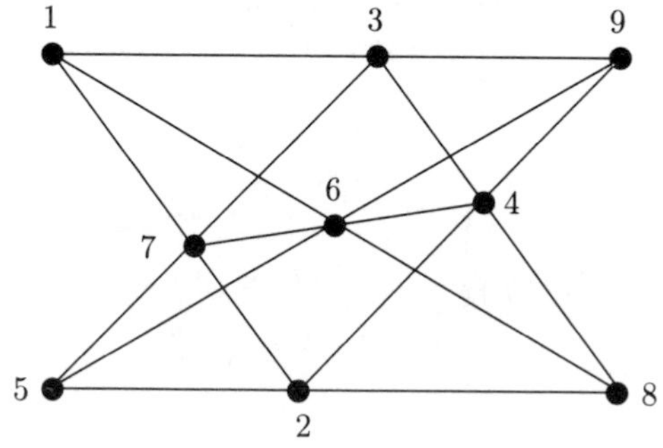

Figure 10-2. The Pappus configuration.

We shall explain this derivation. To this end let us assign homogeneous coordinates $(1:0:0)$ to the point 7, $(0:1:0)$ to the point 8, $(0:0:1)$ to the point 9, and generic coordinates $(x_{i1}:x_{i2}:x_{i3})$ to the points $i = 1,2,3,4,5,6$. The nine collinearities in the Pappus configuration translate precisely into the vanishing of the nine binomials in $in_\omega(I_\mathcal{A})$. For instance, the requirement that the points 1, 3 and 9 are collinear translates into the identity $x_{11}x_{32} - x_{31}x_{12} = 0$. Therefore the combinatorial structure of the nine binomials in $in_\omega(I_\mathcal{A})$ mirrors the combinatorial structure of the Pappus configuration.

Example 8.13 can be derived in a similar manner from the *Vamos matroid*. For details on configurations and matroids see e.g. (Björner et al. 1993).

We shall now present the proof of the Arnold-Korkina-Post-Roelofs-Theorem.

Proof of Theorem 10.2: Let $\mathcal{A} = \{a_1, a_2, a_3\} \subset \mathbf{N}$ and I any $\mathcal{A}$-graded ideal in $k[x_1, x_2, x_3]$. By the *degree* of a monomial $x_1^{i_1}x_2^{i_2}x_3^{i_3}$ we shall always mean the integer $i_1a_1 + i_2a_2 + i_3a_3$.

Case 1: I is generated by monomials. We may assume that x_1 is a non-zerodivisor modulo I. We will show that $I = in_\omega(I_\mathcal{A})$ for the weight vector $\omega \in \mathbf{N}^3$ which is defined as follows:

$$\omega_1 := 0, \quad \omega_2 := min\{\, r : x_3^r \in (I : x_1^\infty)\}, \quad \omega_3 := min\{\, s : x_2^s \in (I : x_1^\infty)\}.$$

Let D be the largest degree (with respect to the $\mathcal{A}$-grading) of a minimal generator of I. Our weights have the property that a monomial $x_1^{\alpha_1}x_2^{\alpha_2}$ (resp. $x_1^{\beta_1}x_3^{\beta_3}$) of degree $\geq D$ lies in I if and only if $\alpha_2 \geq \omega_3$ (resp. $\beta_3 \geq \omega_2$). Consider any vector $\mathbf{u} \in ker(\mathcal{A})$ such that $\mathbf{x}^{\mathbf{u}^-} \notin I$ (and hence $\mathbf{x}^{\mathbf{u}^+} \in I$). It suffices to prove that $\omega \cdot \mathbf{u} > 0$. We can assume that $\mathbf{x}^{\mathbf{u}^+} = x_1^{\alpha_1} x_2^{\alpha_2}$ and $\mathbf{x}^{\mathbf{u}^-} = x_1^{\beta_1} x_3^{\beta_3}$, after swapping the roles of x_2 and x_3 if necessary.

Subcase 1.1: $degree(\mathbf{u}^-) = a_1\beta_1 + a_3\beta_3$ is greater than D. In this case we have $\alpha_2 \geq \omega_3$ and $\beta_3 < \omega_2$. This immediately implies $\omega \cdot \mathbf{u} = \omega_2\alpha_2 - \omega_3\beta_3 > 0$.

Subcase 1.2: $degree(\mathbf{u}^-)$ is arbitrary. We assume that, on the contrary, $\omega_2\alpha_2 \leq \omega_3\beta_3$. By Subcase 1.1 this implies that $x_1^{\alpha_1+D} x_2^{\alpha_2}$ and $x_1^{\beta_1+D} x_3^{\beta_3}$ both lie in I. There exists a unique standard monomial of the same degree, say $x_1^{\gamma_1} x_2^{\gamma_2} x_3^{\gamma_3} \notin I$. Since $x_1^{\beta_1} x_3^{\beta_3}$ is the unique standard monomial of its degree, it follows that $\gamma_1 < D$. This implies $a_2\gamma_2 + a_3\gamma_3 > max\{a_2\alpha_2, a_3\beta_3\}$. The following drawing of the points $(\alpha_2, 0), (0, \beta_3), (\gamma_2, \gamma_3)$ and the linear functionals $(a_2, a_3), (\omega_2, \omega_3)$ shows that $\omega_2\gamma_2 + \omega_3\gamma_3 > min\{\omega_2\alpha_2, \omega_3\beta_3\} = \omega_2\alpha_2$. This is a contradiction to the conclusion of Subcase 1.1.

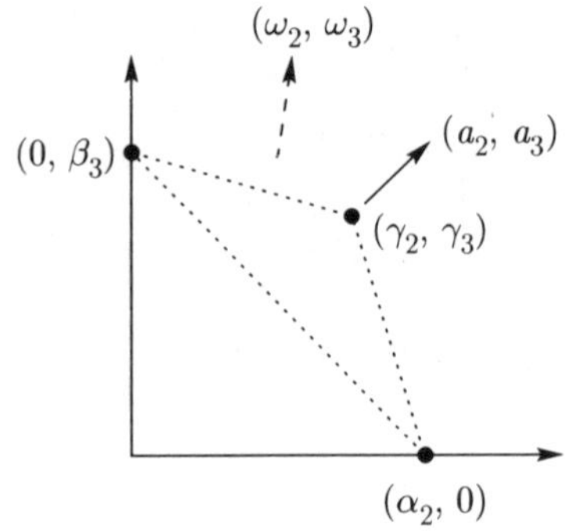

Figure 10-3. A triangle in the proof of Theorem 10.2.

Case 2: I has precisely one minimal generator which is not a monomial. After scaling the variables, we may assume that this generator is $\mathbf{x}^{\mathbf{u}^+} - \mathbf{x}^{\mathbf{u}^-}$. Let $\mathbf{x}^{\mathbf{v}_1}, \ldots, \mathbf{x}^{\mathbf{v}_m}$ be the other (monomial) minimal generators of I. By Case 1, there exist vectors $\omega_1, \omega_2 \in \mathbf{R}^3$ such that

$$\langle \mathbf{x}^{\mathbf{u}^+}, \mathbf{x}^{\mathbf{v}_1}, \ldots, \mathbf{x}^{\mathbf{v}_m} \rangle \subseteq in_{\omega_1}(I_\mathcal{A}) \quad \text{and} \quad \langle \mathbf{x}^{\mathbf{u}^-}, \mathbf{x}^{\mathbf{v}_1}, \ldots, \mathbf{x}^{\mathbf{v}_m} \rangle \subseteq in_{\omega_2}(I_\mathcal{A}).$$

Moreover, we may assume that $in_{\omega_j}(I_\mathcal{A}) = in_{\omega_j}(I)$ for $j = 1, 2$. Let $\mathbf{x}^{\mathbf{c}_{ij}}$ denote the unique monomial not in $in_{\omega_j}(I_\mathcal{A})$ such that $\mathbf{x}^{\mathbf{v}_i} - \mathbf{x}^{\mathbf{c}_{ij}} \in I_\mathcal{A}$, for $i = 1, \ldots, m$ and $j = 1, 2$.

Subcase 2.1: $\mathbf{c}_{i1} = \mathbf{c}_{i2}$ for $i = 1, \ldots, m$. We define $\omega := (\omega_1\mathbf{u}) \cdot \omega_2 - (\omega_2\mathbf{u}) \cdot \omega_1$. Then $\omega \cdot \mathbf{u}^+ = \omega \cdot \mathbf{u}^-$ and $\omega \cdot \mathbf{v}_i > \omega \cdot \mathbf{c}_{i1}$ for all i. This implies that $I \subseteq in_\omega(I_\mathcal{A})$. Since both ideals have the same Hilbert function, we may conclude that $I = in_\omega(I_\mathcal{A})$.

Subcase 2.2: There exists i such that $\mathbf{c}_{i1} \neq \mathbf{c}_{i2}$. Then $\mathbf{x}^{\mathbf{c}_{i1}}$ lies in $in_{\omega_2}(I_\mathcal{A})$ but not in I. This implies that $\mathbf{x}^{\mathbf{v}_i}$ and $\mathbf{x}^{\mathbf{c}_{i1}}$ are relatively prime: otherwise we could factor out a variable from the binomial $\mathbf{x}^{\mathbf{v}_i} - \mathbf{x}^{\mathbf{c}_{i1}} \in I_\mathcal{A}$ and find a proper factor of $\mathbf{x}^{\mathbf{v}_i}$ lying in $in_{\omega_1}(I_\mathcal{A}) = in_{\omega_1}(I)$. This is impossible because $\mathbf{x}^{\mathbf{v}_i}$ is a minimal generator of I, and the reduced Gröbner basis of I with respect to ω_1 consists of monomials and one binomial $\mathbf{x}^{\mathbf{u}^+} - \mathbf{x}^{\mathbf{u}^-}$.

Let $\mathcal{G}_2$ be the reduced Gröbner basis of I with respect to ω_2. The only non-monomial in $\mathcal{G}_2$ is $\mathbf{x}^{\mathbf{u}^-} - \mathbf{x}^{\mathbf{u}^+}$. Since $\mathbf{x}^{\mathbf{c}_{i1}}$ reduces to a multiple of $\mathbf{x}^{\mathbf{c}_{i2}}$ modulo $\mathcal{G}_2$, we have that $\mathbf{x}^{\mathbf{u}^-}$ divides $\mathbf{x}^{\mathbf{c}_{i1}}$. We conclude that $\mathbf{x}^{\mathbf{v}_i}$ and $\mathbf{x}^{\mathbf{u}^-}$ are relatively prime. Applying the same reasoning to the monomial $\mathbf{x}^{\mathbf{c}_{i1}}$, we conclude that $\mathbf{x}^{\mathbf{v}_i}$ and $\mathbf{x}^{\mathbf{u}^-}$ are relatively prime.

The three monomials $\mathbf{x}^{\mathbf{v}_i}$, $\mathbf{x}^{\mathbf{u}^-}$ and $\mathbf{x}^{\mathbf{u}^+}$ are pairwise relatively prime. Since there are only three variables in the ambient polynomial ring, we can write $\mathbf{x}^{\mathbf{v}_i} = x_1^{i_1}$, $\mathbf{x}^{\mathbf{u}^+} = x_2^{i_2}$, and $\mathbf{x}^{\mathbf{u}^-} = x_3^{i_3}$. Note that the integers i_2 and i_3 are relatively prime. We observe that x_2x_3 is a non-zerodivisor modulo I. If not, then $x_2^{mi_2}x_3^{mi_3} \in I$ for some $m \gg 0$, and hence $\langle x_1^{i_1}, x_2^{2mi_2}, x_3^{2mi_2}\rangle \subset I$, which would imply that $k[x_1,x_2,x_3]/I$ is artinian.

We will prove that $I = in_{(1,0,0)}(I_{\mathcal{A}})$. By $\mathcal{A}$-gradedness, it suffices to show that every minimal generator f of $in_{(1,0,0)}(I_{\mathcal{A}})$ lies in I. If f is a binomial, then $f = x_2^{i_2} - x_3^{i_3} \in I$. If f is a monomial, then $f = \mathbf{x}^{\mathbf{u}^+}$ for $\mathbf{u} = \mathbf{u}^+ - \mathbf{u}^- \in ker(\mathcal{A})$ with $1 \in supp(\mathbf{u}^+)$. The relation $supp(\mathbf{u}^-) \subseteq \{2,3\}$ implies $\mathbf{x}^{\mathbf{u}^-} \notin I$. There exists a unique constant $c \in k$ such that $\mathbf{x}^{\mathbf{u}^+} - c \cdot \mathbf{x}^{\mathbf{u}^-} \in I$. If c were non-zero, then $(\mathbf{x}^{\mathbf{u}^-})^{i_1}$ would lie in I, in contradiction to the fact that x_2x_3 is a non-zerodivisor modulo I. Therefore $c = 0$ and we are done.

Case 3: I has at least two minimal generators which are not monomials. After scaling the variables, we can assume that these generators are $\mathbf{x}^{\mathbf{a}^+} - \mathbf{x}^{\mathbf{a}^-}$ and $\mathbf{x}^{\mathbf{b}^+} - \mathbf{x}^{\mathbf{b}^-}$, where $\mathbf{a}$ and $\mathbf{b}$ are linearly independent vectors in $ker(\mathcal{A}) \simeq \mathbf{Z}^2$. If I contains no monomials, then I is torus isomorphic to $I_{\mathcal{A}}$ by Lemma 10.12. Therefore let $\mathbf{x}^{\mathbf{v}} \in I$, and suppose that no proper factor of $\mathbf{x}^{\mathbf{v}}$ lies in I. Choose a monomial $\mathbf{x}^{\mathbf{v}'} \notin I$ such that $\mathbf{x}^{\mathbf{v}} - \mathbf{x}^{\mathbf{v}'} \in I_{\mathcal{A}}$. Clearly, $\mathbf{x}^{\mathbf{v}}$ and $\mathbf{x}^{\mathbf{v}'}$ are relatively prime. Otherwise, if x_i is a common factor, then $\mathbf{x}^{\mathbf{v}'}/x_i - c \cdot \mathbf{x}^{\mathbf{v}}/x_i \in I$ for some $c \in k$, and this implies $\mathbf{x}^{\mathbf{v}'} \in I$.

Suppose there exists a binomial $\mathbf{x}^{\mathbf{u}^+} - \mathbf{x}^{\mathbf{u}^-}$ in I such that $\mathbf{x}^{\mathbf{u}^-}$ divides $\mathbf{x}^{\mathbf{v}'}$. By the same reasoning as in the end of the previous paragraph, we find that $\mathbf{x}^{\mathbf{v}'-\mathbf{u}^-+\mathbf{u}^+}$ is not in I and is therefore relatively prime from $\mathbf{x}^{\mathbf{v}} \in I$. We conclude that $\mathbf{x}^{\mathbf{v}}, \mathbf{x}^{\mathbf{u}^+}$ and $\mathbf{x}^{\mathbf{u}^-}$ are pairwise relatively prime. This means we are in the situation of Subcase 2.2, which has been taken care of already. Therefore we may assume that no such binomial exists.

This assumption implies $\mathbf{x}^{\mathbf{v}'} \notin in_{\prec}(I)$ for every term order $\prec$. Note that we always have $\mathbf{x}^{\mathbf{v}} \in in_{\prec}(I)$. We select term orders $\prec_1, \prec_2, \prec_3, \prec_4$ such that

$$\begin{aligned}
&\mathbf{x}^{\mathbf{a}^+}, \mathbf{x}^{\mathbf{b}^+} \in in_{\prec_1}(I), \quad \mathbf{x}^{\mathbf{a}^-}, \mathbf{x}^{\mathbf{b}^-} \notin in_{\prec_1}(I).\\
&\mathbf{x}^{\mathbf{a}^+}, \mathbf{x}^{\mathbf{b}^-} \in in_{\prec_2}(I), \quad \mathbf{x}^{\mathbf{a}^-}, \mathbf{x}^{\mathbf{b}^+} \notin in_{\prec_1}(I),\\
&\mathbf{x}^{\mathbf{a}^-}, \mathbf{x}^{\mathbf{b}^+} \in in_{\prec_3}(I), \quad \mathbf{x}^{\mathbf{a}^+}, \mathbf{x}^{\mathbf{b}^-} \notin in_{\prec_1}(I),\\
\text{and}\quad &\mathbf{x}^{\mathbf{a}^-}, \mathbf{x}^{\mathbf{b}^-} \in in_{\prec_4}(I), \quad \mathbf{x}^{\mathbf{a}^+}, \mathbf{x}^{\mathbf{b}^+} \notin in_{\prec_1}(I).
\end{aligned}$$

By Case 1, there exist $\omega_i \in \mathbf{R}^3$ such that $in_{\prec_i}(I) = in_{\omega_i}(I_{\mathcal{A}})$ for $i = 1,2,3,4$. This implies

$$\omega_1 \cdot \mathbf{a} > 0\,,\ \omega_2 \cdot \mathbf{a} > 0\,,\ \omega_3 \cdot \mathbf{a} < 0 \text{ and } \omega_4 \cdot \mathbf{a} < 0,$$

$$\omega_1 \cdot \mathbf{b} > 0\,,\ \omega_2 \cdot \mathbf{b} < 0\,,\ \omega_3 \cdot \mathbf{b} > 0 \text{ and } \omega_4 \cdot \mathbf{b} < 0,$$

Since $\{\mathbf{a}, \mathbf{b}\}$ is a basis for the two-dimensional vector space $ker_{\mathbf{R}}(\mathcal{A})$, there exist positive reals $\lambda_i > 0$ such that $\lambda_1\omega_1 + \lambda_2\omega_2 + \lambda_3\omega_3 + \lambda_4\omega_4$ is orthogonal to $ker(\mathcal{A})$. In particular, it is orthogonal to $\mathbf{v} - \mathbf{v}'$, which shows that $\omega_i \cdot \mathbf{v} \leq \omega_i \cdot \mathbf{v}'$ for at least one i. This implies that $\mathbf{x}^{\mathbf{v}'}$ lies in $in_{\prec_i}(I) = in_{\omega_i}(I_{\mathcal{A}})$, which is a contradiction. ∎

Exercises:

(1) Let $n = d + 1$ where $d = dim(\mathcal{A})$. Show that there are precisely three isomorphism types of $\mathcal{A}$-graded algebras and that all three are coherent.

(2) List all $\mathcal{A}$-graded ideals for $\mathcal{A} = \{1, 2, 3\}$.

(3) List all $\mathcal{A}$-graded monomial ideals for $\mathcal{A} = \{1, 3, 4, 7\}$.

(4) Give an algorithm for constructing all $\mathcal{A}$-graded monomial ideals, for an arbitrary configuration $\mathcal{A} \subset \mathbf{N}^d \backslash \{0\}$.

(5) List all polyhedral subdivisions of the configuration in Example 10.11. Does each of them correspond to an $\mathcal{A}$-graded ideal ?

(6) Does there exist a non-coherent $\mathcal{A}$-graded algebra in the case where $\mathcal{A}$ is the vertex set of a regular d-dimensional cube ?

(7) Extend Lemma 10.14 to include all polyhedral subdivisions (not just triangulations) of a unimodular configuration.

Notes:

Arnold (1989) expressed the number of isomorphism classes of $\mathcal{A}$-graded algebras for $\mathcal{A} = \{1, p_2, p_3\}$ in terms of the continued fraction expansion for the rational number p_3/p_2. The extension to the case $\mathcal{A} = \{p_1, p_2, p_3\}$ was given in (Korkina, Post & Roelofs 1995). Our proof of Theorem 10.2 is based on this article.

All results in this chapter are new and not published elsewhere (with the exception of Theorem 10.2, of course). The first example of an infinite family of pairwise non-isomorphic AGA's was constructed by D. Eisenbud (unpublished) for $d = 1, n = 7$. The $n = 4$ example in Theorem 10.4 was found afterwards. Peeva's Theorem 10.13 constitutes a counterexample to a conjecture which I stated after Theorem 10.10 had been found. Theorem 10.15 is an extension of the results in (De Loera 1995a).

CHAPTER 11

Canonical Subalgebra Bases

Toric ideals arise naturally in the study of canonical subalgebra bases. It is the objective of this chapter to explain this connection and to develop an intrinsic Gröbner basis theory for subalgebras of the polynomial ring. The basic idea is to degenerate the algebra generators into monomials and thereby the algebra relations to binomials. Geometrically speaking, we wish to deform an arbitrary parametrically presented variety X into a toric variety. As an application we shall see how this can be accomplished if X is a Grassmann variety.

Let R be a finitely generated subalgebra of the polynomial ring $k[\mathbf{t}] = k[t_1, \ldots, t_d]$. Fix a term order $\prec$ on $k[\mathbf{t}]$. The *initial algebra* $in_\prec(R)$ is the k-vector space spanned by $\{in_\prec(f) : f \in R\}$. A *canonical basis* is a subset $\mathcal{C}$ of R such that $in_\prec(R)$ is generated as a k-algebra by the set of monomials $\{in_\prec(f) : f \in \mathcal{C}\}$. Canonical bases for subalgebras are similar to Gröbner bases with regard to their reduction properties.

Algorithm 11.1. *(The subduction algorithm for a canonical basis $\mathcal{C}$)*
Input: A canonical basis $\mathcal{C}$ for a subalgebra $R \subset k[\mathbf{t}]$. A polynomial $f \in k[\mathbf{t}]$.
Output: An expression of f as a polynomial in the elements of $\mathcal{C}$, provided $f \in R$.

While f is not a constant in k do

1. *Find $f_1, f_2, \ldots, f_r \in \mathcal{C}$, exponents $i_1, i_2, \ldots, i_r \in \mathbf{N}$ and $c \in k^*$ such that*
$$in_\prec(f) \quad = \quad c \cdot in_\prec(f_1)^{i_1} \cdot in_\prec(f_2)^{i_2} \cdots in_\prec(f_r)^{i_r}. \tag{11.1}$$
2. *If no representation (11.1) exists, then output "f does not lie in R" and STOP.*
3. *Otherwise, output $p := c \cdot f_1^{i_1} f_2^{i_2} \cdots f_r^{i_r}$, and replace f by $f - p$.*

Output the constant f.

A nice example of a canonical subalgebra basis is the set of elementary symmetric polynomials $\sigma_1, \ldots, \sigma_d$ in $t_1, \ldots, t_d$. Indeed, the familiar algorithm for expressing symmetric polynomials in terms of $\sigma_1, \ldots, \sigma_d$ is precisely Algorithm 11.1 in this case.

The main difference between Gröbner bases for ideals and canonical bases for subalgebras is that the initial algebra $in_\prec(R)$ need not be finitely generated. If $in_\prec(R)$ is not finitely generated, then there is no finite canonical basis for R with respect to $\prec$. The following example appears in (Göbel 1995).

Example 11.2. *(The invariants of the alternating group have an infinite canonical basis)* Let $d = 3$ and let $R = k[t_1, t_2, t_3]^{A_3}$ be the subalgebra of polynomials which are invariant under the cyclic permutation $t_1 \mapsto t_2, t_2 \mapsto t_3, t_3 \mapsto t_1$. It has four minimal generators:

$$R \quad = \quad k\big[\, t_1 + t_2 + t_3,\; t_1 t_2 + t_1 t_3 + t_2 t_3,\; t_1 t_2 t_3,\; (t_1 - t_2)(t_1 - t_3)(t_2 - t_3) \,\big].$$

Let $\prec$ be the lexicographic term order with $t_1 \succ t_2 \succ t_3$. If f is any invariant in R and $in_\prec(f) = t_1^{i_1}t_2^{i_2}t_3^{i_3}$, then it is easy to see that either $i_1 \geq i_2 \geq i_3$ or $i_1 > i_3 \geq i_2$. Suppose that $in_\prec(R)$ is finitely generated. Among the generators consider the subset $\{t_1^{a_1}t_3^{b_1}, t_1^{a_2}t_3^{b_2}, \cdots, t_1^{a_s}t_3^{b_s}\}$ of those generators which do not contain the variable t_2. There exists a constant $C > 1$ such that $a_i \geq C \cdot b_i$ for $i = 1, \ldots, s$. Choose any integer $d > \frac{1}{C-1}$, so that $d + 1 < C \cdot d$. We consider the A_3-invariant polynomial

$$g \quad := \quad \underline{t_1^{d+1}t_3^d} + t_1^d t_2^{d+1} + t_2^d t_3^{d+1} \quad \in \quad R.$$

The underlined initial term must lie in the semigroup generated by the $t_1^{a_i}t_3^{b_i}$. This implies that the vector $(d+1, d)$ lies in the planar convex cone spanned by the vectors (a_i, b_i). Hence it also satisfies the linear inequality $d+1 \geq C \cdot d$, a contradiction. This shows that $in_\prec(R)$ is not finitely generated. ■

It is an important open problem to find good criteria which guarantee finite generation for $in_\prec(R)$. In what follows we consider mainly the case where $in_\prec(R)$ is finitely generated. (Most of our discussion about canonical bases, however, would extend to the infinite case.)

Fix a set of polynomials $\mathcal{F} = \{f_1, f_2, \ldots, f_n\}$ in $k[\mathbf{t}] = k[t_1, \ldots, t_d]$, let $R = k[\mathcal{F}]$ be the subalgebra they generate, and fix a term order $\prec$ on $k[\mathbf{t}]$. Suppose $in_\prec(f_i) = \mathbf{t}^{\mathbf{a}_i}$ and let $\mathcal{A} = \{\mathbf{a}_1, \mathbf{a}_2, \ldots, \mathbf{a}_n\} \subset \mathbf{N}^d$. We shall give a criterion for deciding whether $\mathcal{F}$ is a canonical basis for R with respect to $\prec$. To this end we introduce the new polynomial ring $k[\mathbf{x}] = k[x_1, x_2, \ldots, x_n]$. Consider the k-algebra epimorphism from $k[\mathbf{x}]$ onto R defined by $x_i \mapsto f_i$, and let I denote its kernel. Similarly, consider the map from $k[\mathbf{x}]$ onto $in_\prec(R)$ defined by $x_i \mapsto in_\prec(f_i) = \mathbf{t}^{\mathbf{a}_i}$. The kernel of this map is the toric ideal $I_\mathcal{A}$.

Let $\omega \in \mathbf{R}^d$ be any weight vector which represents the term order $\prec$ for the polynomials in $\mathcal{F}$. If we consider $\mathcal{A}$ as a $d \times n$-matrix, with transpose $\mathcal{A}^T$, then $\mathcal{A}^T\omega$ is a vector in $\mathbf{R}^n$. We can use it as the weight vector for forming an initial ideal of $I \subset k[\mathbf{x}]$. However, the initial ideal $in_{\mathcal{A}^T\omega}(I)$ is usually not a monomial ideal. This is explained by the fact that the vector $\mathcal{A}^T\omega$ is *not* a generic vector in $\mathbf{R}^n$, even if ω is generic in $\mathbf{R}^d$.

Lemma 11.3. *For any set $\mathcal{F} \subset k[\mathbf{t}]$, the initial ideal $in_{\mathcal{A}^T\omega}(I)$ is contained in the toric ideal $I_\mathcal{A}$.*

Proof. Let $p(\mathbf{x}) = \sum c_\mathbf{u}\mathbf{x}^\mathbf{u}$ be an element of the ideal I. This means that

$$p(f_1(\mathbf{t}), \ldots, f_n(\mathbf{t})) \quad = \quad \sum c_\mathbf{u} f_1(\mathbf{t})^{u_1} \cdots f_n(\mathbf{t})^{u_n} \quad \in \quad k[\mathbf{t}]$$

is the zero polynomial. When expanding this sum, the terms of highest ω-order must cancel. The ω-order of $f_1(\mathbf{t})^{u_1} \cdots f_n(\mathbf{t})^{u_n}$ equals the ω-order of

$$in_\omega(f_1)(\mathbf{t})^{u_1} \cdots in_\omega(f_n)(\mathbf{t})^{u_n} \;=\; \mathbf{t}^{u_1\mathbf{a}_1} \cdots \mathbf{t}^{u_n\mathbf{a}_n},$$

which is the inner product of $\mathcal{A}^T\omega$ with $\mathbf{u} = (u_1, \ldots, u_n)$. Therefore the sum of the highest terms in the expansion $\sum c_\mathbf{u} f_1(\mathbf{t})^{u_1} \cdots f_n(\mathbf{t})^{u_n}$ equals

$$\big[in_{\mathcal{A}^T\omega}(p)\big]\big(in_\omega(f_1), \ldots, in_\omega(f_n)\big) = \big[in_{\mathcal{A}^T\omega}(p)\big]\big(\mathbf{t}^{\mathbf{a}_1}, \ldots, \mathbf{t}^{\mathbf{a}_n}\big) = 0. \qquad (11.2)$$

We conclude that $in_{\mathcal{A}^T\omega}(p)$ lies in $I_\mathcal{A}$. ■

The reverse inclusion to Lemma 11.3 is our criterion for canonical bases.

Theorem 11.4. *The set $\mathcal{F} \subset k[\mathbf{t}]$ is a canonical basis if and only if $in_{\mathcal{A}^T\omega}(I) = I_{\mathcal{A}}$.*

Proof: We first show the "only-if" direction. Suppose $\mathcal{F}$ is a canonical basis, and let q be any binomial in $I_{\mathcal{A}}$. We consider the polynomial $q(f_1, \ldots, f_n)$ in $k[\mathbf{t}]$. By the canonical basis property and using Algorithm 11.1, we can find $r \in k[\mathbf{x}]$ such that $q(f_1, \ldots, f_n) = r(f_1, \ldots, f_n)$ and $q = in_{\mathcal{A}^T\omega}(q - r)$. This proves $q \in in_{\mathcal{A}^T\omega}(I)$ as desired.

The proof of the "if" direction is by contradiction. Suppose that $in_{\mathcal{A}^T\omega}(I) = I_{\mathcal{A}}$ but $\mathcal{F}$ is not a canonical basis. Then there exists $p \in k[\mathbf{x}]$ such that

$$in_\omega\big(p(f_1, \ldots, f_n)\big) \;\notin\; k\big[in_\omega(f_1), \ldots, in_\omega(f_n)\big]. \tag{11.3}$$

We may assume that p is minimal with respect to the partial term order defined by $\mathcal{A}^T\omega$. In order for (11.3) to hold, it must be the case that the terms $\mathbf{t}^{\mathbf{u}}$ of highest order in the expansion of $p(f_1(\mathbf{t}), \ldots, f_n(\mathbf{t}))$ all cancel. As in the proof of Lemma 11.3, this implies (11.2) and therefore $in_{\mathcal{A}^T\omega}(p) \in I_{\mathcal{A}} = in_{\mathcal{A}^T\omega}(I)$. There exists a polynomial $q \in I$ such that $in_{\mathcal{A}^T\omega}(p) = in_{\mathcal{A}^T\omega}(q)$. The initial form of $p - q$ with respect to $\mathcal{A}^T\omega$ is therefore smaller than that of p. However, since $q \in I$, we have $p(f_1, \ldots, f_n) = (p - q)(f_1, \ldots, f_n)$, so that $p - q$ shares the property (11.3) with p. This is a contradiction to the minimality in our choice of p, and the proof is complete. ∎

In order to apply the criterion in Theorem 11.4 one has to compute generators for the toric ideal $I_{\mathcal{A}}$, but one need *not* do this for I. Instead one uses Algorithm 11.1.

Corollary 11.5. *Let $\{p_1, \ldots, p_s\}$ be generators of the toric ideal $I_{\mathcal{A}}$. Then $\mathcal{F}$ is a canonical basis if and only if Algorithm 11.1 reduces $p_i(f_1, \ldots, f_n)$ to a constant for all $i \in \{1, \ldots, s\}$.*

Proof: The only-if direction is obvious: if Algorithm 11.1 gets stuck with a non-constant polynomial while reducing $p_i(f_1, \ldots, f_n)$, then $\mathcal{F}$ fails to be a canonical basis. For the if-direction we assume that Algorithm 11.1 does reduce $p_i(f_1, \ldots, f_n)$ to a constant. The output generated by that reduction gives a polynomial $q_i(\mathbf{x}) = \sum_{i=1}^{t} c_i \mathbf{x}^{\mathbf{u}_i}$ whose terms form a strictly descending sequence in the partial term order $\succ$ defined by $\mathcal{A}^T\omega$ on $k[\mathbf{x}]$:

$$in_{\mathcal{A}^T\omega}(p_i) \;\succ\; \mathbf{x}^{\mathbf{u}_1} \;\succ\; \mathbf{x}^{\mathbf{u}_2} \;\succ\; \cdots \;\succ\; \mathbf{x}^{\mathbf{u}_t}. \tag{11.4}$$

By construction, we have $p_i(f_1, \ldots, f_n) = q_i(f_1, \ldots, f_n)$, and hence $p_i - q_i \in I$. Property (11.4) implies $p_i = in_{\mathcal{A}^T\omega}(p_i - q_i) \in in_{\mathcal{A}^T\omega}(I)$. This holds for all $i \in \{1, \ldots, s\}$, and, in view of Lemma 11.3, we conclude

$$I_{\mathcal{A}} \;=\; \langle p_1, p_2, \ldots, p_s \rangle \;=\; in_{\mathcal{A}^T\omega}(I).$$

Using Theorem 11.4, this completes the proof. ∎

The initial ideal $in_{\mathcal{A}^T\omega}(I)$ occurring in Theorem 11.4 is not a monomial ideal (yet). It is natural to ask how its different Gröbner bases enter the overall picture.

Corollary 11.6. *Using notation as above, suppose that $\mathcal{F}$ is a canonical basis.*

(1) Every reduced Gröbner basis $\mathcal{G}$ of $I_{\mathcal{A}}$ lifts to a reduced Gröbner basis $\mathcal{H}$ of I, i.e., the elements of $\mathcal{G}$ are the initial forms (with respect to $\mathcal{A}^T\omega$) of the elements of $\mathcal{H}$.

(2) Every regular triangulation of $\mathcal{A}$ is an initial complex of the ideal I.

(3) The state polytope of $I_{\mathcal{A}}$ is a face of the state polytope of the homogenization of I.

Proof: Let $\mathcal{G}$ be the reduced Gröbner basis of $I_{\mathcal{A}}$ with respect to a term order $\prec$, and let $\mathcal{H}$ be the reduced Gröbner basis of I with respect to $\prec_{\mathcal{A}^T\omega}$. By Proposition 1.8, we have

$$in_{\prec}(I_{\mathcal{A}}) \quad = \quad in_{\prec}\big(in_{\mathcal{A}^T\omega}(I)\big) \quad = \quad in_{\prec_{\mathcal{A}^T\omega}}(I). \tag{11.5}$$

This implies $in_{\prec}(in_{\mathcal{A}^T\omega}(\mathcal{H})) = in_{\prec}(\mathcal{G})$. Since all trailing terms of elements in $\mathcal{H}$ are $\prec$-standard, we conclude that $in_{\mathcal{A}^T\omega}(\mathcal{H}) = \mathcal{G}$. This proves (1).

Part (2) follows directly from (11.5) and Theorem 8.3. Let I_{homog} denote the homogenization of I. We extend $\mathcal{A}^T\omega$ to a partial term order for I_{homog} by making the homogenizing variable reverse lexicographically smallest. For part (3) we use the following consequence of Lemma 2.6:

$$State(I_{\mathcal{A}}) \quad = \quad State(in_{\mathcal{A}^T\omega}(I_{homog})) \quad = \quad face_{\mathcal{A}^T\omega}\big(State(I_{homog})\big).$$

Here we had to replace I by I_{homog} because the Gröbner fan of I need not be complete (in which case the polytope $State(I)$ is not defined). However, the toric ideal $I_{\mathcal{A}}$ is positively graded since the columns of $\mathcal{A}$ are non-zero non-negative vectors, so that $State(I_{\mathcal{A}})$ is always a well-defined polytope. ∎

Corollary 11.5 gives rise to a simple completion algorithm for computing a canonical basis from any finite generating set $\mathcal{F}$, provided $in_{\prec}(k[\mathcal{F}])$ is finitely generated. Namely, if there exists a minimal generator p_i of $I_{\mathcal{A}}$ such that Algorithm 11.1 reduces $p_i(f_1,\ldots,f_n)$ to a non-constant polynomial $q_i(f_1,\ldots,f_n)$, then simply add $q_i(f_1,\ldots,f_n)$ to the set $\mathcal{F}$ and proceed. Just as in the case of the Buchberger algorithm for ideals, this completion procedure can be made more efficient by auto-reductions and other more clever strategies.

Example 11.7. *(The algebra generated by the 2×2-minors of a generic 3×3-matrix)* Consider a 3×3-matrix of indeterminates (t_{ij}) and let $\mathcal{F}$ be the set of its 2×2-minors,

$$\begin{array}{lll} f_1 = \underline{t_{11}t_{22}} - t_{12}t_{21}, & f_2 = \underline{t_{11}t_{23}} - t_{13}t_{21}, & f_3 = \underline{t_{12}t_{23}} - t_{13}t_{22}, \\ f_4 = \underline{t_{11}t_{32}} - t_{12}t_{31}, & f_5 = \underline{t_{11}t_{33}} - t_{13}t_{31}, & f_6 = \underline{t_{12}t_{33}} - t_{13}t_{32}, \\ f_7 = \underline{t_{21}t_{32}} - t_{22}t_{31}, & f_8 = \underline{t_{21}t_{33}} - t_{23}t_{31}, & f_9 = \underline{t_{22}t_{33}} - t_{23}t_{32}. \end{array} \tag{11.6}$$

We fix a term order $\prec$ which selects the main diagonal term to be the initial term, for each minor of (t_{ij}). The ideal of algebraic relations among the underlined initial terms has only two generators:

$$I_{\mathcal{A}} \quad = \quad \langle\, x_4x_8 - x_5x_7,\; x_2x_6 - x_3x_5 \,\rangle.$$

Under the substitution $x_i \mapsto f_i(\mathbf{t})$ we get

$$f_{10} := f_4 f_8 - f_5 f_7 = t_{31} \cdot det(t_{ij}) \qquad \text{and} \qquad f_{11} := f_2 f_6 - f_3 f_5 = t_{13} \cdot det(t_{ij}),$$

where $det(t_{ij})$ is the determinant of the given 3×3-matrix. Neither of the initial terms

$$in_{\prec}(f_{10}) = t_{11}t_{22}t_{31}t_{33} \qquad \text{or} \qquad in_{\prec}(f_{11}) = t_{11}t_{13}t_{22}t_{33}$$

lies in the subalgebra generated by the nine underlined monomials in (11.6). We enlarge the generating set to $\mathcal{F}' := \mathcal{F} \cup \{f_{10}, f_{11}\}$. The corresponding toric ideal remains unchanged:

$$I_{\mathcal{A}'} = \langle\, x_4x_8 - x_5x_7,\, x_2x_6 - x_3x_5 \,\rangle,$$

since the variable t_{31} (resp. t_{13}) occurs only in $in_{\prec}(f_{10})$ (resp. $in_{\prec}(f_{11})$) but in no other initial term. This implies that $\mathcal{F}'$ is a canonical basis for its subalgebra $k[\mathcal{F}] = k[\mathcal{F}']$. ∎

Example 11.7 stands in a certain contrast to the next result which concerns maximal minors. Consider a matrix of indeterminates (t_{ij}) of format $r \times s$, where $r \leq s$. Let R be the subalgebra of $k[\mathbf{t}] = k[t_{11}, \ldots, t_{rs}]$ generated by the $r \times r$-minors of (t_{ij}). Its projective spectrum $Proj(R)$ is the *Grassmann variety* $Grass_{r,s}$ of r-dimensional linear subspaces in an s-dimensional vector space, presented in its usual *Plücker embedding*. A term order on $k[\mathbf{t}]$ is called *diagonal* if the main diagonal term is the initial term for each $r \times r$-minor.

Theorem 11.8. *The set of $r \times r$-minors of an $r \times s$-matrix of indeterminates is a canonical basis for the subalgebra they generate, with respect to any diagonal term order on $k[\mathbf{t}]$.*

Proof: See Theorem 3.2.9 in (Sturmfels 1993a). ∎

We associate a new variable $[i_1 i_2 \cdots i_r]$ to the $r \times r$-minor with column indices $i_1 < i_2 < \cdots < i_r$. Thus the polynomial ring $k[\mathbf{x}]$ is generated by these $\binom{s}{r}$ *brackets.* Let $I_{r,s}$ denote the ideal in $k[\mathbf{x}] = k\big[\cdots, [i_1 \cdots i_r], \cdots\big]$ generated by the algebraic relations among the $r \times r$-minors. The ideal $I_{r,s}$ is called the *Grassmann-Plücker ideal.* Theorem 11.8 is a consequence of the classical *straightening algorithm* for $I_{r,s}$.

Let $\mathbf{e}_{ij}$ denote the unit vector in $\mathbf{N}^{r \times s}$ corresponding to the variable t_{ij}. The vector configuration associated with the diagonal initial monomials $t_{1i_1} t_{2i_2} \cdots t_{ri_r}$ equals

$$\mathcal{A}_{r,s} = \big\{\, \mathbf{e}_{1i_1} + \mathbf{e}_{2i_2} + \cdots + \mathbf{e}_{ri_r} \;:\; 1 \leq i_1 < i_2 < \cdots < i_r \leq s \,\big\}. \qquad (11.7)$$

The toric ideal $I_{\mathcal{A}_{r,s}}$ is the kernel of the map

$$k\big[\cdots, [i_1 i_2 \cdots i_r], \cdots\big] \;\rightarrow\; k[t_{11}, t_{12}, \ldots, t_{rs}], \quad [i_1 i_2 \cdots i_r] \;\mapsto\; t_{1i_1} t_{2i_2} \cdots t_{ri_r}. \qquad (11.8)$$

Example 11.9. *(The Grassmann variety of lines in projective 4-space)*
In the special case $r = 2$ and $s = 5$ we shall prove Theorem 11.8 by applying

the criterion in Theorem 11.4. The algebra of interest is generated by the ten 2×2-minors of the matrix

$$\mathbf{t} \quad = \quad \begin{pmatrix} t_{11} & t_{12} & t_{13} & t_{14} & t_{15} \\ t_{21} & t_{22} & t_{23} & t_{24} & t_{25} \end{pmatrix}$$

A diagonal term order is given, for instance, by the following weight matrix

$$\omega \quad = \quad \begin{pmatrix} 1 & 2 & 3 & 4 & 5 \\ 1 & 4 & 9 & 16 & 25 \end{pmatrix}.$$

The ten diagonal initial terms generate the toric algebra of the configuration (11.7):

$$k[\mathcal{A}_{2,5}] = k[\,t_{11}t_{22},\, t_{11}t_{23},\, t_{11}t_{24},\, t_{11}t_{25},\, t_{12}t_{23},\, t_{12}t_{24},\, t_{12}t_{25},\, t_{13}t_{24},\, t_{13}t_{25},\, t_{14}t_{25}\,].$$

We consider the map (11.8) from the free polynomial ring

$$k[\mathbf{x}] \quad = \quad k\big[\,[12],[13],[14],[15],[23],[24],[25],[34],[35],[45]\,\big].$$

onto $k[\mathcal{A}_{2,5}]$. Its kernel is the toric ideal $I_{\mathcal{A}_{2,5}}$. This ideal is generated by the five binomials

$$[14][23]-[13][24],\; [15][23]-[13][25],\, [15][24]-[14][25],$$
$$[15][34]-[14][35],\; [25][34]-[24][35].$$

To establish Theorem 11.8, we must verify the inclusion $I_{\mathcal{A}} \subseteq in_{\mathcal{A}^T\omega}(I_{2,5})$. The induced weight vector $\mathcal{A}^T\omega$ has the entry $i + j^2$ in the coordinate indexed by $[ij]$. Each of the following five *Plücker relations* lies in the kernel $I_{2,5}$ of the canonical epimorphism $k[\mathbf{x}] \to R$:

$$\begin{aligned} &\underline{[14][23]-[13][24]} + [12][34], \\ &\underline{[15][23]-[13][25]} + [12][35], \\ &\underline{[15][24]-[14][25]} + [12][45], \\ &\underline{[15][34]-[14][35]} + [13][45], \\ &\underline{[25][34]-[24][35]} + [23][45]. \end{aligned} \tag{11.9}$$

The underlined initial forms selected by the weight $\mathcal{A}^T\omega$ coincide with the generators. This completes the proof of Theorem 11.8 in the special case $r = 2$ and $s = 5$. ■

Theorem 11.8 and Corollary 11.6 have the following geometric implications. By a *toric deformation* we mean a flat deformation using a one-parameter subgroup of the torus $(k^*)^n$.

Proposition 11.10.
(a) There exists a toric deformation taking the Grassmann variety $Grass_{r,s}$ into the projective toric variety defined by the configuration $\mathcal{A}_{r,s}$.
(b) Every initial ideal of the toric ideal $I_{\mathcal{A}_{r,s}}$ is an initial ideal of the Grassmann-Plücker ideal $I_{r,s}$.

(c) *Every regular triangulation of $\mathcal{A}_{r,s}$ is an initial complex of the Grassmann variety.*

(d) *The state polytope of $\mathcal{A}_{r,s}$ is a face of the state polytope of the Grassmann variety.*

Remark 11.11. *(Digression into algebraic combinatorics)*
The classical straightening algorithm for the Grassmann-Plücker ideal is a special case of the Gröbner bases arising from Proposition 11.10 (b). The best route to seeing this is the following detour through the land of algebraic combinatorics: The set $\mathcal{A}_{r,s}$ is (affinely isomorphic to) the vertex set of the *order polytope* of the product of two chains $[r] \times [s-r]$. The *distributive lattice* $\mathcal{L} = J\big([r] \times [s-r]\big)$ is (isomorphic to) the poset of brackets in the coordinatewise order. We depict this lattice for $r = 2, s = 5$ (Example 11.9):

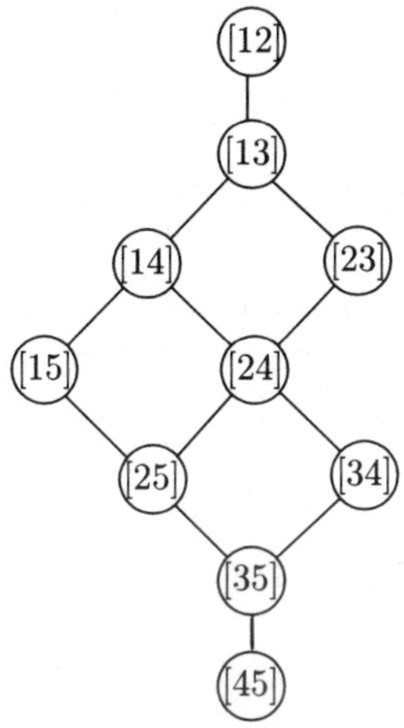

Figure 11-1. The distributive lattice $J\big([2] \times [3]\big)$.

The toric ideal $I_\mathcal{A}$ is generated by the binomials $x \cdot y - (x \vee y) \cdot (x \wedge y)$ where $x, y \in \mathcal{L}$ stand for brackets and $\vee$ and $\wedge$ are the lattice operations. We define a term order on $k[\mathbf{x}] = k[\mathcal{L}]$ as follows: first sort the variables by any linear extension of $\mathcal{L}$ and then sort monomials by the reverse lexicographic order. The initial ideal of $I_\mathcal{A}$ coincides with the initial ideal of the Grassmann-Plücker ideal $I_{r,s}$. It is generated by all products $x \cdot y$ of incomparable elements in $\mathcal{L}$. The initial complex is the *chain complex* of $\mathcal{L}$, which is known to be a regular triangulation of the order polytope of $[r] \times [s-r]$. ∎

Not every Gröbner bases of $I_{r,s}$ arises from a Gröbner basis of $I_{\mathcal{A}_{r,s}}$, i.e., the converse of Proposition 11.10 (b) does not hold. We will demonstrate this in the case $r = 2$.

Corollary 11.12. *The toric ideal $I_{\mathcal{A}_{2,s}}$ has the following properties:*

(a) *The set of circuits equals the universal Gröbner basis:*

$$[i_1j_1][i_2j_2]\cdots[i_\nu j_\nu] - [i_2j_1][i_3j_2]\cdots[i_1j_\nu],$$
$$(i_1, i_2 < j_1 \text{ and } i_2, i_3 < j_2 \text{ and } \ldots \text{ and } i_\nu, i_1 < j_\nu).$$

(b) *All initial ideals of $I_{\mathcal{A}_{2,s}}$ are square-free.*

Proof: The configuration $\{\mathbf{e}_{1i} + \mathbf{e}_{2j} \,:\, 1 \le i, j \le s\}$ is isomorphic to the vertex set of the product of simplices $\Delta_{s-1} \times \Delta_{s-1}$; see Example 5.1. It is unimodular by Exercise (9) in Chapter 8. Therefore its subset $\mathcal{A}_{2,s} = \{\mathbf{e}_{1i}+\mathbf{e}_{2j} \,:\, 1 \le i < j \le s\}$ is unimodular as well. Using Remark 8.10 this proves the assertion (b). To prove (a) we note that the circuits of $\Delta_{s-1} \times \Delta_{s-1}$ are identified with the circuits in the complete bipartite graph $K_{s,s}$. What is listed in (a) is the subset of circuits whose support lies in $\mathcal{A}_{2,s}$. To complete the proof we use Proposition 8.11. ∎

The Grassmann-Plücker ideal $I_{2,6}$ has initial ideals which are not square-free. For example, choose the weight vector

$$\omega = (9, 56, 82, 40, 86, 95, 55, 85, 88, 88, 39, 46, 10, 26, 62)$$

for the $\binom{6}{2} = 15$ brackets $[ij]$ in the usual lexicographic order. We get an initial monomial ideal $in_\omega(I_{2,6})$ among whose minimal generators there is $[15][23]^2[46]$. By Corollary 11.12 (b), the monomial ideal $in_\omega(I_{2,6})$ is not an initial ideal of $I_{\mathcal{A}_{2,6}}$.

We also remark that statement (b) does not hold for $r \ge 3$; for instance, $\mathcal{A}_{3,6}$ contains the vertices of a regular 3-cube as a subset, hence it has a regular triangulation one of whose simplices does not have unit volume, hence it has an initial ideal that is not square-free.

Here is an open problem: *What is the maximum degree $F(r,s)$ appearing in any reduced Gröbner basis for the Grassmann-Plücker ideal $I_{r,s}$?* In view of Proposition 11.10 (b) and Proposition 4.11, the number $F(r,s)$ is bounded below by the maximum degree of any circuit of the configuration $\mathcal{A}_{r,s}$. For instance, Corollary 11.12 (a) implies

$$F(2,s) \quad \ge \quad s-2. \tag{11.10}$$

For an explicit proof of this inequality we may consider the degree $s-2$ circuit

$$[13][24][35]\cdots[s-2,s] \;-\; [23][34][45]\cdots[1,s]. \tag{11.11}$$

To construct an initial ideal of the Grassmann-Plücker ideal which has one of the monomials in (11.11) as a minimal generator, start with the *Vandermonde weights* $[ij] \mapsto i + j^2$ and break ties using an elimination order for the variables appearing in (11.11).

Returning to our general discussion, we assume that $\mathcal{F} = \{f_1, \ldots, f_n\} \subset k[t_1, \ldots, t_d]$ is a canonical basis with respect to $\omega \in \mathbf{R}^d$. We shall present an *intrinsic* Gröbner basis theory for the "canonically presented" subalgebra $k[\mathcal{F}]$. Let J be any ideal in $k[\mathcal{F}]$. The *initial ideal* of J is the following ideal in the initial algebra $k[\mathcal{A}] = in_\omega(k[\mathcal{F}]) = k[in_\omega(\mathcal{F})]$:

$$in_\omega(J) \quad := \quad \langle\, in_\omega(f) \,:\, f \in J \,\rangle. \tag{11.12}$$

An important difference from "classical" Gröbner basis theory is this: the original ideal J lies in $k[\mathcal{F}]$, but the initial ideal $in_\omega(J)$ lies in a different ring, namely, it lies in the toric ring $k[\mathcal{A}]$. A subset $\mathcal{G}$ of J is a *Gröbner basis* with respect to ω if $in_\omega(J)$ is generated by $\{in_\omega(g) \,:\, g \in \mathcal{G}\}$. If this set minimally generates $in_\omega(J)$, then $\mathcal{G}$ is a *minimal* Gröbner basis. From now on we shall always assume that ω is a *term order for* J, that is, $in_\omega(J)$ is generated by monomials $\mathbf{t}^{\mathbf{c}}$, where $\mathbf{c} \in \mathbf{N}\mathcal{A}$. A minimal Gröbner basis $\mathcal{G}$ of J is *reduced* if none of the trailing terms $\mathbf{t}^{\mathbf{b}}$ appearing in any $g \in \mathcal{G}$ are contained in $in_\omega(J)$.

Lemma 11.13. *Every ideal $J \subset k[\mathcal{F}]$ possesses a unique finite reduced Gröbner basis.*

Proof: We first show uniqueness. Suppose $\mathcal{G}$ and $\mathcal{G}'$ are two distinct reduced Gröbner bases of J. There exist $g \in \mathcal{G}$ and $g' \in \mathcal{G}'$ such that $g \neq g'$ but $in_\omega(g) = in_\omega(g')$. The polynomial $g - g'$ lies in $J \backslash \{0\}$ but none of its terms lies in $in_\omega(J)$. This is a contradiction.

We next show existence. Since $k[\mathcal{A}]$ is Noetherian and positively graded, the monomial ideal $in_\omega(J)$ has a unique finite minimal generating set of the form $\{\mathbf{t}^{\mathbf{c}_1}, \mathbf{t}^{\mathbf{c}_2}, \dots, \mathbf{t}^{\mathbf{c}_r}\}$. For each $i \in \{1, \dots, r\}$ there exists an element $g_i \in J$ such that $in_\omega(g_i) = \mathbf{t}^{\mathbf{c}_i}$. The set $\{g_1, g_2, \dots, g_r\}$ is a minimal Gröbner basis. To replace it by the reduced Gröbner basis, we successively reduce each element g_j by the complementary set $\{g_1, \dots, g_{j-1}, g_{j+1}, \dots, g_r\}$ using Subroutine 11.14 below. This reduction process terminates because ω defines a Noetherian order on the monomials $\{\mathbf{t}^{\mathbf{b}} : \mathbf{b} \in \mathbf{N}\mathcal{A}\} \subset k[\mathbf{t}]$. ∎

Subroutine 11.14. *(Reduction inside a canonically presented subalgebra $k[\mathcal{F}]$)*
Input: A polynomial $p \in k[\mathcal{F}]$, a finite subset $\mathcal{G} \subset k[\mathcal{F}]$, a term order ω on $k[\mathbf{t}]$.
Output: A normal form for p modulo $\mathcal{G}$ with respect to ω.

While the polynomial p contains a term $\mathbf{t}^{\mathbf{b}}$ which lies in the ideal $\langle in_\omega(\mathcal{G})\rangle$ do:

1. *Let $g \in \mathcal{G}$ such that $\mathbf{t}^{\mathbf{b}}$ lies in the principal ideal generated by $in_\omega(g) = \mathbf{t}^{\mathbf{d}}$.*
2. *Find an integral vector $\lambda = (\lambda_1, \dots, \lambda_n)$ in the fiber $\pi^{-1}(\mathbf{b} - \mathbf{d})$.*
3. *Replace p by $p - g \cdot \prod_{i=1}^n f_i^{\lambda_i}$.*

The vector λ chosen in Step 2 has the property

$$in_\omega\Big(\prod_{i=1}^n f_i^{\lambda_i}\Big) \;=\; \prod_{i=1}^n (in_\omega(f_i))^{\lambda_i} \;=\; \prod_{i=1}^n (\mathbf{t}^{\mathbf{a}_i})^{\lambda_i} \;=\; \mathbf{t}^{\pi(\lambda)} \;=\; \mathbf{t}^{\mathbf{b}-\mathbf{d}}.$$

Therefore the subtraction in Step 3 does indeed cancel the term $\mathbf{t}^{\mathbf{b}}$ from p. The main difficulty of Subroutine 11.14 lies in the fact that testing membership in a monomial ideal of $k[\mathcal{A}]$ amounts to solving a disjunction of integer programming feasibility problems. Incidentally, we encounter a similar difficulty already in Step 1 of Algorithm 11.1.

Remark 11.15. *In the toric ring $k[\mathcal{A}]$ we have*

$$\mathbf{t}^{\mathbf{b}} \in \langle \mathbf{t}^{\mathbf{c}_1}, \dots, \mathbf{t}^{\mathbf{c}_r}\rangle \qquad \text{if and only if} \qquad \bigcup_{i=1}^r \pi^{-1}(\mathbf{b} - \mathbf{c}_i) \neq \emptyset.$$

To generalize Buchberger's criterion, we introduce the module of syzygies

$$Syz(\mathbf{t}^{\mathbf{c}_1}, \dots, \mathbf{t}^{\mathbf{c}_r}) \;=\; \big\{ (h_1, \dots, h_r) \in k[\mathcal{A}]^r \,:\, h_1\mathbf{t}^{\mathbf{c}_1} + \cdots + h_r\mathbf{t}^{\mathbf{c}_r} = 0 \big\}. \tag{11.13}$$

Theorem 11.16. *Let $\mathcal{G} = \{g_1, \dots, g_r\} \subset k[\mathcal{F}]$, and let $\mathcal{H}$ be any subset of $k[\mathcal{F}]^r$ such that $\{(in_\omega(h_1), \dots, in_\omega(h_r)) : (h_1, \dots, h_r) \in \mathcal{H}\}$ generates the $k[\mathcal{A}]$-module $Syz(in_\omega(g_1), \dots, in_\omega(g_r))$. Then $\mathcal{G}$ is a Gröbner basis for the ideal $\langle \mathcal{G}\rangle$ with respect to ω if and only if, for every $h = (h_1, \dots, h_r) \in \mathcal{H}$, the polynomial $h_1g_1 + \cdots + h_rg_r$ reduces to zero modulo $\mathcal{G}$ via Subroutine 11.14.*

Proof: See Theorem 4.9 in (Miller 1996). ∎

Algorithm 11.17. *(Intrinsic Buchberger for a canonically presented subalgebra)*
Input: A generating set $\mathcal{S}$ of an ideal $J \subset k[\mathcal{F}]$, a term order $\omega \in \mathbf{R}^d$.
Output: The reduced Gröbner basis $\mathcal{G}$ of J with respect to ω.

1. *Let $\mathcal{S} = \{s_1, \ldots, s_r\}$ where $in_\omega(s_i) = \mathbf{t}^{\mathbf{c}_i}$.*
2. *Compute a finite generating set $\mathcal{M}$ for $Syz(\mathbf{t}^{\mathbf{c}_1}, \ldots, \mathbf{t}^{\mathbf{c}_r}) \subset k[\mathcal{A}]^r$ (Subroutine 11.18).*
3. *Set* `newguys` $:= \emptyset$.
4. *For each $m = (m_1, \ldots, m_r) \in \mathcal{M}$ do:*
   4.1. *Find $h = (h_1, \ldots, h_r) \in k[\mathcal{F}]^r$ such that $in_\omega(h_j) = m_j$ for $j = 1, \ldots, r$.*
   4.2. *Compute the normal form $\overline{h}$ of $h_1 s_1 + \cdots + h_r s_r$ modulo $\mathcal{S}$ (Subroutine 11.14).*
   4.3. `newguys` $:=$ `newguys` $\cup \{\overline{h}\}$.
5. *If* `newguys` $\neq \{0\}$ *then set* $\mathcal{S} := \mathcal{S} \cup$ `newguys`$\backslash\{0\}$, *and return to Step 1.*
6. *If* `newguys` $= \{0\}$ *then*
   6.1. *Compute the auto-reduction $\mathcal{G}$ of $\mathcal{S}$ (by applying Subroutine 11.14 to reduce s modulo $\mathcal{S}\backslash\{s\}$, for all $s \in \mathcal{S}$).*
   6.2. *Output $\mathcal{G}$.*

This leaves us with the problem of how to compute generators for the syzygy module (11.13). We shall present two subroutines (11.18 and 11.21) for performing this task. We write $\mathbf{e}_i$ for the standard basis vectors in the free module $k[\mathcal{A}]^r$.

Subroutine 11.18. *(Computing the syzygies on some monomials in a toric ring)*
Input: A vector of monomials $(\mathbf{t}^{\mathbf{c}_1}, \ldots, \mathbf{t}^{\mathbf{c}_r}) \in k[\mathcal{A}]^r$.
Output: A finite generating set for $Syz(\mathbf{t}^{\mathbf{c}_1}, \ldots, \mathbf{t}^{\mathbf{c}_r}) \subset k[\mathcal{A}]^r$.

1. *Find $\mathbf{u}_i \in \pi^{-1}(\mathbf{c}_i)$ for $i = 1, \ldots, r$.*
2. *Let $\mathcal{S} \subset k[\mathbf{x}]^r$ be any generating set for the syzygies on $(\mathbf{x}^{\mathbf{u}_1}, \ldots, \mathbf{x}^{\mathbf{u}_r})$, for instance, the usual S-pairs. Apply the toric homomorphism $x_i \mapsto \mathbf{t}^{\mathbf{a}_i}$ to $\mathcal{S}$ and output the result.*
3. *Compute a reduced Gröbner basis (in the ordinary sense) for the ideal intersection*

$$\langle \mathbf{x}^{\mathbf{u}_1}, \ldots, \mathbf{x}^{\mathbf{u}_r} \rangle \cap I_\mathcal{A} \quad \text{in} \quad k[\mathbf{x}] = k[x_1, \ldots, x_n]. \tag{11.14}$$

4. *For each element $g = g(\mathbf{x})$ in the reduced Gröbner basis of (11.14) do:*
   4.1. *Write g in the form $g(\mathbf{x}) = \mathbf{x}^{\mathbf{v}} \cdot \mathbf{x}^{\mathbf{u}_i} - \mathbf{x}^{\mathbf{w}} \cdot \mathbf{x}^{\mathbf{u}_j}$, where $i, j \in \{1, \ldots, r\}$.*
   4.2. *Output the syzygy $\mathbf{t}^{\pi(\mathbf{v})} \cdot \mathbf{e}_i - \mathbf{t}^{\pi(\mathbf{w})} \cdot \mathbf{e}_j$.*

Example 11.19. Here is a simple example of a syzygy module which is not generated by S-pairs. Take $d = 2, n = 3$ and $\mathcal{A} = \{(2,0), (1,1), (0,2)\}$, so that $k[\mathcal{A}] = k[t_1^2, t_1t_2, t_2^2] = k[x_1, x_2, x_3]/\langle x_1x_3 - x_2^2 \rangle$. Then $Syz(t_1^2, t_1t_2)$ is minimally generated by $(t_1t_2, -t_1^2)$ and $(t_2^2, -t_1t_2)$. The first syzygy is found in Step 2 and the second is found in Step 4.2. ∎

The correctness of Subroutine 11.18 is the content of Proposition 4.10 in (Miller 1996). In Step 3.1 we are making implicitly the claim that every reduced Gröbner basis of (11.14) consists of binomials $\mathbf{x}^{\mathbf{v}} \cdot \mathbf{x}^{\mathbf{u}_i} - \mathbf{x}^{\mathbf{w}} \cdot \mathbf{x}^{\mathbf{u}_j}$. To prove this claim, we recall the standard algorithm for computing ideal intersections (Cox, Little & O'Shea 1992; §4.3, Theorem 11): Introduce a new variable z, form the ideal $B := \langle (1-z) \cdot \mathbf{x}^{\mathbf{u}_1}, \ldots, (1-z) \cdot \mathbf{x}^{\mathbf{u}_r} \rangle + z \cdot I_\mathcal{A}$ in $k[\mathbf{x}, z]$, and then compute the elimination ideal $B \cap k[\mathbf{x}]$. Our claim follows because B is a binomial ideal and the Buchberger algorithm is "binomial-friendly". See Corollary 1.7 in (Eisenbud & Sturmfels 1996) for a more general result.

These considerations imply the following toric generalization of the familiar fact that S-pairs suffice to generate all syzygies (Cox, Little & O'Shea 1992; §4.3, Proposition 8).

Corollary 11.20. *Syzygies on monomials in $k[\mathcal{A}]$ are generated by pairwise syzygies:*

$$Syz\,(\sum_{i=1}^{r} \mathbf{t}^{\mathbf{c}_i} \cdot \mathbf{e}_i\,) \quad = \quad \sum_{1 \leq i < j \leq r} Syz\,(\,\mathbf{t}^{\mathbf{c}_i} \cdot \mathbf{e}_i \,+\, \mathbf{t}^{\mathbf{c}_j} \cdot \mathbf{e}_j\,).$$

Corollary 11.20 reduces the computation of the syzygy module (11.13) to the special case $r = 2$. Hence we could also use the following subroutine for Step 2 in Algorithm 11.17.

Subroutine 11.21. *(Computing syzygies on a pair of monomials in a toric ring)*
Input: Two monomials $\mathbf{t}^{\mathbf{c}}$ and $\mathbf{t}^{\mathbf{d}}$ in $k[\mathcal{A}]$.
Output: A finite generating set for $Syz(\mathbf{t}^{\mathbf{c}}, \mathbf{t}^{\mathbf{d}}) \subset k[\mathcal{A}]^2$.

1. *Form the toric ideal $I_{\mathcal{A}\cup\{\mathbf{c}-\mathbf{d}\}} \subset k[x_1, \ldots, x_n, z]$, where z is mapped to $\mathbf{t}^{\mathbf{c}-\mathbf{d}}$.*
2. *Compute the reduced Gröbner basis $\mathcal{G}$ for $I_{\mathcal{A}\cup\{\mathbf{c}-\mathbf{d}\}}$ with respect to any elimination order $z \succ \{x_1, \ldots, x_n\}$.*
3. *For each binomial in $\mathcal{G}$ which contains z linearly, such as $\mathbf{x}^{\mathbf{u}} \cdot z - \mathbf{x}^{\mathbf{v}}$, output the corresponding syzygy $(\mathbf{t}^{\pi(\mathbf{u})}, -\mathbf{t}^{\pi(\mathbf{v})})$.*

Proof of correctness: It follows immediately from the construction that each output pair $(\mathbf{t}^{\pi(\mathbf{u})}, -\mathbf{t}^{\pi(\mathbf{v})})$ is a syzygy of $(\mathbf{t}^{\mathbf{c}}, \mathbf{t}^{\mathbf{d}})$. Conversely, every minimal syzygy can be written as a pair $(\mathbf{t}^{\pi(\mathbf{u}')}, -\mathbf{t}^{\pi(\mathbf{v}')})$ such that $\mathbf{x}^{\mathbf{u}'} \cdot z - \mathbf{x}^{\mathbf{v}'}$ lies in $I_{\mathcal{A}\cup\{\mathbf{c}-\mathbf{d}\}}$. The Gröbner basis property of $\mathcal{G}$ implies that there exists a binomial $\mathbf{x}^{\mathbf{u}} \cdot z - \mathbf{x}^{\mathbf{v}} \in \mathcal{G}$ such that $\mathbf{x}^{\mathbf{u}}$ divides $\mathbf{x}^{\mathbf{u}'}$ and $\mathbf{x}^{\mathbf{v}'} - \mathbf{x}^{\mathbf{v}+\mathbf{u}'-\mathbf{u}}$ lies in $I_{\mathcal{A}}$. The given syzygy therefore equals $\mathbf{t}^{\pi(\mathbf{u}'-\mathbf{u})} \cdot (\mathbf{t}^{\pi(\mathbf{u})}, -\mathbf{t}^{\pi(\mathbf{v})})$. ∎

Example 11.22. The minimal number of generators of $Syz(\mathbf{t}^{\mathbf{c}}, \mathbf{t}^{\mathbf{d}})$ cannot be bounded by a function in n, d and $\mathcal{A}$. Let $d{=}3, n{=}7$ and

$$k[\mathcal{A}] = k[t_1t_2t_3, t_1^2t_2, t_1t_2^2, t_1^2t_3, t_1t_3^2, t_2^2t_3, t_2t_3^2].$$

It can be shown that the minimal generators of $Syz\big(t_1^{2i}t_2^{2i}t_3^{2i},\, t_1^{3i+2}t_2t_3^{3i}\,\big)$ include $2i + 2$ syzygies of total degree $i + 1$, for $i \geq 1$. ∎

Any bound must therefore involve the degrees of $\mathbf{c}$ and $\mathbf{d}$. Here is such a bound.

Theorem 11.23. *Let $D(\,\cdot\,)$ be defined as in Theorem 4.7. Then the total degree (in $k[\mathbf{x}]^r$) of any minimal generators of the syzygy module (11.13) is bounded above by*

$$max_{1\leq i<j\leq r}\ (d+1)\cdot(n+1-d)\cdot D\big(\mathcal{A}\cup\{\mathbf{c}_i-\mathbf{c}_j\}\big).$$

Proof: This follows from Corollary 11.20, Subroutine 11.21 and Theorem 4.7. ∎

One disadvantage of the intrinsic Buchberger Algorithm 11.17 is that – at present – it is not available in any computer algebra system. However, there is an "extrinsic" method for simulating Algorithm 11.17, which is easy to run in the currently available Gröbner bases programs.

Algorithm 11.24. *(Extrinsic computation of intrinsic Gröbner bases)*
Input: Generators for an ideal J in $k[\mathcal{F}]$ and a term order ω on $k[\mathbf{t}]$.
Output: A Gröbner basis for J with respect to ω.

1. *Let I denote the kernel of the canonical epimorphism*
$$\phi : k[\mathbf{x}] \to k[\mathcal{F}],\ x_i \mapsto f_i(\mathbf{t}).$$
2. *For each generator of J choose a preimage, and let $\overline{J} \subset k[\mathbf{x}]$ be the ideal they generate.*
3. *Compute the reduced Gröbner basis $\mathcal{G}$ of the ideal $I + \overline{J}$ with respect to any term order refining the weight vector $\mathcal{A}^T\omega$.*
4. *Output its image $\phi(\mathcal{G}) = \{\phi(g) : g \in \mathcal{G}\}$ in $k[\mathcal{F}]$.*

Proof of correctness: Clearly, $\phi(\mathcal{G}) \subset J$. The fact that the Gröbner basis $\mathcal{G}$ is reduced implies

$$in_\omega(\phi(g)) \quad = \quad in_{\mathcal{A}^T\omega}(g)(\mathbf{t}^{\mathbf{a}_1},\ldots,\mathbf{t}^{\mathbf{a}_n}) \qquad \text{for all} \quad g \in \mathcal{G}. \tag{11.15}$$

We must show that $\{\, in_{\mathcal{A}^T\omega}(g)(\mathbf{t}^{\mathbf{a}_1},\ldots,\mathbf{t}^{\mathbf{a}_n}) \ :\ g \in \mathcal{G}\}$ generates $in_\omega(J)$. Let $h \in J \subset k[\mathcal{F}]$. By the canonical basis property of $\mathcal{F} = \{f_1,\ldots,f_n\}$, there exists $p \in k[\mathbf{x}]$ such that $h = p(f_1,\ldots,f_n)$ and $in_\omega(h) = in_{\mathcal{A}^T\omega}(p)(\mathbf{t}^{\mathbf{a}_1},\ldots,\mathbf{t}^{\mathbf{a}_n}) \neq 0$. Since $p \in I + \overline{J}$, its initial form $in_{\mathcal{A}^T\omega}(p)$ lies in the ideal $\langle in_{\mathcal{A}^T\omega}(\mathcal{G})\rangle$ in $k[\mathbf{x}]$. This implies that $in_\omega(h)$ lies in (11.15), as desired. ∎

We discuss a geometric example in the Grassmann variety of lines in projective 4-space.

Example 11.25. Consider the canonical basis in Theorem 11.8 for $r = 2, s = 5$. Here $\mathcal{F} = \{[12],[13],\ldots,[45]\}$ is the set of 2×2-minors of a 2×5-matrix of indeterminates (t_{ij}), and ω is the weight with coordinates $\omega_{ij} = j^i$. Consider the ideal $J = \langle g_1, g_2\rangle$, where

$$g_1 \ = \ ([12]-[23])\cdot([23]-[34]) \quad \text{and} \quad g_2 \ = \ ([13]+[14]+[15])\cdot([15]+[25]+[35]).$$

The corresponding subvariety of the Grassmann variety consists of all lines in P^4 which meet the following two pairs of codimension 2 subspaces: $\{x_2 = x_1 + x_3 = 0\} \cup \{x_3 = x_2 + x_4 = 0\}$ and $\{x_1 = x_3 + x_4 + x_5 = 0\} \cup \{x_5 = x_1 + x_2 + x_3 = 0\}$. We wish to compute the Gröbner basis of J intrinsically in $k[\mathcal{F}]$. The initial terms of the generators are

$$in_\omega(g_1) \ = \ t_{12}t_{13}t_{23}t_{24} \quad \text{and} \quad in_\omega(g_2) \ = \ t_{11}t_{13}t_{25}^2.$$

Using Subroutine 11.18 or 11.21, we compute the following minimal generating set for the syzygy module $Syz(t_{12}t_{13}t_{23}t_{24}, t_{11}t_{13}t_{25}^2)$:

$$\{\,(t_{11}t_{13}t_{25}^2, -t_{12}t_{23}t_{13}t_{24}),\ (t_{11}^2t_{25}^2, -t_{11}t_{23}t_{12}t_{24}),\ (t_{11}t_{12}t_{25}^2, -t_{12}^2t_{23}t_{24})\}.$$

Following Step 4.1 of Algorithm 11.17 we express each of these six monomials as an initial form; for instance, $t_{11}t_{13}t_{25}^2 = in_\omega([15][35])$. In Step 4.2 we form the corresponding linear combinations in J:

$$\begin{aligned} g_3 &:= [15][35] \cdot g_1 - [23][34] \cdot g_2, \\ g_4 &:= [15]^2 \cdot g_1 - [13][24] \cdot g_2, \\ g_5 &:= [15][25] \cdot g_1 - [23][24] \cdot g_2. \end{aligned}$$

The polynomial g_3 reduces to zero modulo $\{g_1, g_2\}$. The initial terms of g_4 and g_5 are in normal form modulo $\{g_1, g_2\}$. Another run through Algorithm 11.17 confirms that $\{g_1, g_2, g_4, g_5\}$ is a minimal Gröbner basis for J. However, to arrive at the reduced Gröbner basis we must further reduce g_4 and g_5 modulo $\{g_1, g_2\}$ in Step 6.1.

We remark that already for this small example the extrinsic computation is quite redundant: the reduced Gröbner basis in Step 3 of Algorithm 11.24 contains 15 elements.

Exercises:

(1) Give an example of a subalgebra $k[\mathcal{F}]$ of $k[\mathbf{x}]$ and two term orders $\prec_1$ and $\prec_2$ such that $in_{\prec_1}(k[\mathcal{F}])$ is finitely generated but $in_{\prec_2}(k[\mathcal{F}])$ is not finitely generated.

(2) Let $d = 5, n = 6$ and let $\mathcal{F}$ be the set of 2×2-minors of the matrix

$$\begin{pmatrix} t_1 & t_2 & t_3 & t_4 \\ t_2 & t_3 & t_4 & t_5 \end{pmatrix}.$$

Compute a canonical basis for the subalgebra $k[\mathcal{F}]$.

(3) Compute the state polytope of the Grassmann variety $Grass_{2,5}$. Show that the converse of Proposition 11.12 (b) holds for $r = 2, s = 5$: every initial ideal of the Grassmann-Plücker ideal $I_{2,5}$ is an initial ideal of the toric ideal $I_{\mathcal{A}_{2,5}}$.

(4) Compute the universal Gröbner basis of the toric ideal $I_{\mathcal{A}_{3,6}}$. Use your answer to give a lower bound on $F(3, 6)$.

(5) Show that the intersection of two principal ideals $\langle \mathbf{t}^\mathbf{b} \rangle$ and $\langle \mathbf{t}^\mathbf{c} \rangle$ in a toric ring $k[\mathcal{A}]$ can have arbitrarily many minimal generators.

(6) The ring of symmetric polynomials $k[x_1, \ldots, x_n]^{S_n}$ is canonically presented by the set $\mathcal{F}$ of elementary symmetric functions. Implement Algorithm 11.17 in this case.

Notes:

The concept of canonical bases was introduced independently by Kapur & Madlener (1989) and Robbiano & Sweedler (1990). Further properties and applications of canonical bases were studied in (Ollivier 1991). Ollivier's results include some remarkable connections between the algebraic operation of taking integral closure and convexity properties of the initial algebra. Miller (1996) extends the theory of canonical bases to polynomial rings over general base rings. While Theorem 11.8

appears in (Sturmfels 1993a, §3.2), the point of view presented in Example 11.9 and Proposition 11.10 is new. Toric varieties arising from lattices (as in Remark 11.11) are studied in (Wagner 1996). Our discussion of Gröbner bases inside canonically presented subalgebras is derived from results in (Ollivier 1991) and (Miller 1996). An intrinsic Gröbner basis theory for flag varieties in generic coordinates can be found in (Rippel 1994).

CHAPTER 12

Generators, Betti Numbers and Localizations

This chapter is organized into four sections. Their general theme is "advanced techniques" for computing with toric ideals. We present two algorithms for finding generators and Gröbner bases of $I_{\mathcal{A}}$, and we introduce truncated Gröbner bases with respect to the grading by the semigroup $\mathbf{N}\mathcal{A}$. The generators and higher syzygies of toric ideals are characterized in terms of certain simplicial complexes, and, finally, the technique of localization in integer programming is discussed.

12.A. Computing generators and Gröbner bases

A familiar method for computing generators and Gröbner bases of the toric ideal $I_{\mathcal{A}}$ is Algorithm 4.5. Unfortunately, that algorithm is too slow. It cannot handle non-trivial problems from the domains of application discussed in Chapter 5. In view of the sensitivity of Buchberger's algorithm to the number of variables involved, it is much better to use an algorithm which operates entirely in $k[x_1, \ldots, x_n]$ rather than in the auxiliary polynomial ring $k[t_0, t_1, \ldots, t_d, x_1, \ldots, x_n]$. In what follows we present two such algorithms. These two algorithms appear to be of comparable efficiency.

We recall the definition of *ideal quotients.* If f is a polynomial in $k[\mathbf{x}]$ and $J \subset k[\mathbf{x}]$ is an ideal, then the following two subsets of $k[\mathbf{x}]$ are again ideals:

$$\begin{aligned} (J : f) &= \{ g \in k[\mathbf{x}] \, : \, fg \in J \}, \\ (J : f^\infty) &= \{ g \in k[\mathbf{x}] \, : \, f^r g \in J \text{ for some } r \in \mathbf{N} \}. \end{aligned} \qquad (12.1)$$

A basic formula involving ideal quotients is $(I : fg) = ((I : f) : g)$. A general method for computing Gröbner bases of the ideals in (12.1) from generators of J can be found in Section 4.4 of (Cox, Little & O'Shea 1992). If J is a homogeneous ideal and f is one of the variables, say, $f = x_n$, then the algorithm of choice is furnished by the following lemma.

Lemma 12.1. *Fix the graded reverse lexicographic term order induced by $x_1 > \cdots > x_n$, and let $\mathcal{G}$ be the reduced Gröbner basis of a homogeneous ideal $J \subset k[\mathbf{x}]$. Then the set*

$$\mathcal{G}' \;=\; \{ f \in \mathcal{G} : x_n \text{ does not divide} f \} \cup \{ f/x_n : f \in \mathcal{G} \text{ and } x_n \text{ divides} f \}$$

is a Gröbner basis of $(J : x_n)$. A Gröbner basis of $(J : x_n^\infty)$ is obtained by dividing each element $f \in \mathcal{G}$ by the highest power of x_n that divides f.

Proof: We will show that $\mathcal{G}'$ is a Gröbner basis for $(J : x_n)$. The proof of the second assertion about $(J : x_n^\infty)$ is analogous. Let $g \in (J : x_n)$. Then $in(x_n \cdot g) = x_n \cdot in(g)$ is divisible by $in(f)$ for some $f \in \mathcal{G}$. Our choice of term order guarantees that x_n

divides f if and only if x_n divides $in(f)$. If this is the case, then f/x_n lies in $\mathcal{G}'$ and $in(f/x_n)$ divides $in(g)$. If x_n does not divide f, then f lies in $\mathcal{G}'$ and $in(f)$ divides $in(g)$. In either case $in(g)$ lies in the ideal generated by the initial terms of $\mathcal{G}'$. ∎

The term order used in Lemma 12.1 makes sense whenever the ideal J is homogeneous with respect to some positive grading $deg(x_i) = d_i > 0$. By iterating the Gröbner basis computation n times with respect to different reverse lexicographic term orders, that is, by applying Lemma 12.1 one variable at a time, one can compute the ideal quotient

$$(J : (x_1 x_2 \cdots x_n)^\infty) \quad = \quad ((\cdots (J : x_1^\infty) : x_2^\infty) \cdots) : x_n^\infty). \qquad (12.2)$$

This ideal consists of all polynomials $f \in k[\mathbf{x}]$ such that $f \cdot m \in J$ for some monomial m.

In what follows we assume that $\mathcal{A} = \{\mathbf{a}_1, \ldots, \mathbf{a}_n\}$ is a subset of $\mathbf{N}^d \backslash \{0\}$, so that the toric ideal $I_\mathcal{A}$ is positively graded and has the Gröbner region $\mathbf{R}^n$. Let $ker(\mathcal{A}) \subset \mathbf{Z}^n$ denote the integer kernel of the $d \times n$-matrix with column vectors $\mathbf{a}_i$. With any subset $\mathcal{C}$ of the lattice $ker(\mathcal{A})$ we associate a subideal of $I_\mathcal{A}$:

$$J_\mathcal{C} \quad := \quad \langle \mathbf{x}^{\mathbf{v}^+} - \mathbf{x}^{\mathbf{v}^-} \; : \mathbf{v} \in \mathcal{C} \rangle. \qquad (12.3)$$

Clearly, this ideal is endowed with the same $\mathbf{N}\mathcal{A}$-grading as $I_\mathcal{A}$.

Lemma 12.2. *A subset $\mathcal{C}$ spans the lattice $ker(\mathcal{A})$ if and only if*

$$(J_\mathcal{C} : (x_1 \cdots x_n)^\infty) \quad = \quad I_\mathcal{A}.$$

Proof: It suffices to show this for a finite subset $\mathcal{C} = \{\mathbf{v}_1, \ldots, \mathbf{v}_r\} \subset ker(\mathcal{A})$. For the only-if direction suppose that $\mathbf{u} \in ker(\mathcal{A})$ can be written as $\mathbf{u} = \sum_{i=1}^r \lambda_i \mathbf{v}_i$ with λ_i integers. This implies the following identity of rational functions:

$$\frac{\mathbf{x}^{\mathbf{u}^+}}{\mathbf{x}^{\mathbf{u}^-}} - 1 \quad = \quad \prod_{i=1}^r \left(\frac{\mathbf{x}^{\mathbf{v}_i^+}}{\mathbf{x}^{\mathbf{v}_i^-}}\right)^{\lambda_i} - 1.$$

By clearing denominators we get an identity in $k[\mathbf{x}]$ which shows that a monomial multiple of $\mathbf{x}^{\mathbf{u}^+} - \mathbf{x}^{\mathbf{u}^-}$ lies in $J_\mathcal{C}$. For the converse suppose that $\mathbf{x}^\mathbf{a} \cdot (\mathbf{x}^{\mathbf{u}^+} - \mathbf{x}^{\mathbf{u}^-})$ lies in $J_\mathcal{C}$. By the proof of Theorem 5.3, we can connect the lattice points $\mathbf{a} + \mathbf{u}^+$ and $\mathbf{a} + \mathbf{u}^-$ by a sequence of moves from $\mathcal{C}$. Hence the vector $\mathbf{u} = \mathbf{u}^+ - \mathbf{u}^-$ is an integer linear combination of $\mathcal{C}$. ∎

The two lemmas stated give rise to the following algorithm.

Algorithm 12.3. *(Computing a Gröbner basis of a toric ideal)*
1. Find any lattice basis $\mathcal{B}$ for $ker(\mathcal{A})$.
2. *(Optional)* Replace $\mathcal{B}$ by a lattice basis $\mathcal{B}_{red}$ which is *reduced* in the sense of Lovász.
3. Let $J_0 := \langle \mathbf{x}^{\mathbf{u}_+} - \mathbf{x}^{\mathbf{u}_-} \; : \mathbf{u} \in \mathcal{B}_{red} \rangle$.
4. For $i = 1, 2, \ldots, n$: Compute $J_i := (J_{i-1} : x_i^\infty)$ using Lemma 12.1, that is, by making x_i the reverse lexicographically cheapest variable.
5. Compute the reduced Gröbner basis of $J_n = I_\mathcal{A}$ for the desired term order.

A few comments are in place. The computation of a lattice basis for $ker(\mathcal{A})$ can be done using the *Hermite normal form* algorithm for integer matrices. For details see (Schrijver 1986, §5.3). The *reduced lattice basis* $\mathcal{B}_{red}$ in Step 2 consists of vectors with relatively small integer entries. An implementation of basis reduction is available in MAPLE under the command `lattice`. The underlying theory is explained in (Schrijver 1986, §6.2).

Computational experience with Algorithm 12.3 using the system GRIN is reported in (Hosten & Sturmfels 1995). The subroutines called upon in Steps 1 and 2 are polynomial time algorithms. Their running time is negligible relative to the subsequent Gröbner basis computations in Steps 4 and 5. Step 4 involves n Gröbner basis computations. But these are relatively short and easy calculations, especially when compared with Algorithm 4.5.

We next present a different algorithm, due to Di Biase and Urbanke (1995), for computing a generating set of the toric ideal $I_{\mathcal{A}}$. It is based on Lemma 12.4 and Proposition 12.5 below. In these two assertions we no longer assume that $I_{\mathcal{A}}$ is positively graded.

Lemma 12.4. *Let $\mathcal{C}$ be a spanning set for the lattice $ker(\mathcal{A})$ such that one of the vectors $\mathbf{u} \in \mathcal{C}$ has all coordinates positive. Then the ideal $J_{\mathcal{C}}$ is equal to the toric ideal $I_{\mathcal{A}}$.*

Proof: The binomial $\mathbf{x}^{\mathbf{u}} - 1$ lies in $J_{\mathcal{C}}$. It shows that all variables x_i are invertible modulo $J_{\mathcal{C}}$. This is equivalent to $J_{\mathcal{C}} = \big(J_{\mathcal{C}} : (x_1 x_2 \cdots x_n)^\infty\big)$. Now apply Lemma 12.2. ∎

Let $\mathcal{A}_i$ denote the configuration $\mathcal{A}$ with $\mathbf{a}_i$ replaced by its negative $-\mathbf{a}_i$. In what follows we use $\mathbf{x}^{\mathbf{u}}, \mathbf{x}^{\mathbf{v}}, \mathbf{x}^{\mathbf{u}_j}, \mathbf{x}^{\mathbf{v}_j}$ to denote monomials which do not contain the variable x_i. Then a binomial $x_i^r \mathbf{x}^{\mathbf{u}} - \mathbf{x}^{\mathbf{v}}$ lies in $I_{\mathcal{A}_i}$ if and only if $\mathbf{x}^{\mathbf{u}} - \mathbf{x}^{\mathbf{v}} x_i^r$ lies in $I_{\mathcal{A}}$. Let $\prec$ be any term order on $k[\mathbf{x}]$ which eliminates x_i, that is, all monomials containing x_i are higher than those not containing x_i.

Proposition 12.5. *Let $\mathcal{G}_i = \{x_i^{r_j} \mathbf{x}^{\mathbf{u}_j} - \mathbf{x}^{\mathbf{v}_j} : j = 1, 2, \ldots, m\}$ be a Gröbner basis for $I_{\mathcal{A}_i}$ with respect to $\prec$. Then $\mathcal{G} = \{\mathbf{x}^{\mathbf{u}_j} - x_i^{r_j} \mathbf{x}^{\mathbf{v}_j} : j = 1, 2, \ldots, m\}$ is a generating set for $I_{\mathcal{A}}$.*

Proof: Let $\mathbf{x}^{\mathbf{u}} - x_i^r \mathbf{x}^{\mathbf{v}}$ be a binomial in $I_{\mathcal{A}}$. Clearly, $I_{\mathcal{A}}$ is generated by binomials of this form. Then $x_i^r \mathbf{x}^{\mathbf{u}} - \mathbf{x}^{\mathbf{v}}$ lies in $I_{\mathcal{A}_i}$ and therefore reduces to zero modulo the Gröbner basis $\mathcal{G}_i$. Let $f_1, \ldots, f_m$ be the polynomial coefficients picked up in this zero reduction:

$$x_i^r \mathbf{x}^{\mathbf{u}} - \mathbf{x}^{\mathbf{v}} \quad = \quad \sum_{j=1}^{m} f_j(x_1, \ldots, x_n) \cdot (x_i^{r_j} \mathbf{x}^{\mathbf{u}_j} - \mathbf{x}^{\mathbf{v}_j}). \tag{12.4}$$

The elimination property of the term order $\prec$ implies that the variable x_i occurs with degree at most $r - r_j$ in f_j. (In particular, we have $f_j = 0$ whenever $r < r_j$.) Now replace x_i by $1/x_i$ in (12.4) and thereafter multiply both sides by x_i^r. We obtain the identity

$$\mathbf{x}^{\mathbf{u}} - x_i^r \mathbf{x}^{\mathbf{v}} \quad = \quad \sum_{j=1}^{m} x_i^{r - r_j} \cdot f_j(x_1, \ldots, x_{i-1}, \frac{1}{x_i}, x_{i+1}, \ldots, x_n) \cdot (\mathbf{x}^{\mathbf{u}_j} - x_i^{r_j} \mathbf{x}^{\mathbf{v}_j}).$$

This identity shows that the left hand side is a *polynomial* linear combination of the elements in $\mathcal{G}$. We conclude that $I_{\mathcal{A}} = \langle \mathcal{G} \rangle$, as desired. ∎

Algorithm 12.6. *(Computing a generating set of a toric ideal)*

1. Choose a subset $\{i_1, \ldots, i_r\}$ of $\{1, \ldots, n\}$ such that the kernel of the matrix $\mathcal{A}_{i_1 i_2 \cdots i_r} := (\cdots(\mathcal{A}_{i_1})_{i_2}\cdots)_{i_r}$ contains a strictly positive vector.
2. Find a lattice basis $\mathcal{B}$ for $ker(\mathcal{A}_{i_1 i_2 \cdots i_r})$ which contains a strictly positive vector.
3. The ideal $I_{\mathcal{A}_{i_1 i_2 \cdots i_r}}$ is generated by $\{\mathbf{x}^{\mathbf{u}_+} - \mathbf{x}^{\mathbf{u}_-} : \mathbf{u} \in \mathcal{B}\}$ (by Lemma 12.4).
4. Let $\ell := r$.
5. While $\ell \geq 1$ do
   5.1. Choose a term order which eliminates the variable x_{i_ℓ}.
   5.2. Compute a Gröbner basis $\mathcal{G}_{i_1 i_2 \cdots i_\ell}$ for $I_{\mathcal{A}_{i_1 i_2 \cdots i_\ell}}$.
   5.3. Flip the variable x_{i_ℓ} as in Proposition 12.5 to get generators for $I_{\mathcal{A}_{i_1 i_2 \cdots i_{\ell-1}}}$.
   5.4. $\ell := \ell - 1$
6. Output the resulting generating set for $I_{\mathcal{A}}$.

The number of Gröbner basis computations in Step 5 is the cardinality of the set $\{i_1, \ldots, i_r\}$. Therefore it would be best to use a smallest such "inversion set". The problem of how to find one is best studied in the context of oriented matroids (Björner et.al. 1993). In Step 2 it is advantageous to use a reduced basis, just as in Algorithm 12.3.

Example 12.7. Let $d = 4, n = 8$ and consider the matrix

$$\mathcal{A} \quad = \quad \begin{pmatrix} 1 & 2 & 3 & 4 & 0 & 1 & 4 & 5 \\ 2 & 3 & 4 & 1 & 1 & 4 & 5 & 0 \\ 3 & 4 & 1 & 2 & 4 & 5 & 0 & 1 \\ 4 & 1 & 2 & 3 & 5 & 0 & 1 & 4 \end{pmatrix}.$$

Each column of $\mathcal{A}$ has the same sum, namely 10. This implies that the toric ideal $I_\mathcal{A}$ is homogeneous in the usual grading. The reduced Gröbner basis of $I_\mathcal{A}$ with respect to the lexicographic term order given by $x_1 > x_2 > x_3 > x_4 > x_5 > x_6 > x_7 > x_8$ equals

$$\begin{aligned} \mathcal{G} = \{\, & x_1^3 - x_3 x_5^2,\; x_1^2 x_7 - x_3^2 x_5,\; x_1 x_3 - x_5 x_7,\; x_1 x_7^2 - x_3^3,\; x_2^3 - x_4 x_6^2, \\ & x_2^2 x_8 - x_4^2 x_6,\; x_2 x_4 - x_6 x_8,\; x_2 x_8^2 - x_4^3,\; x_3^4 - x_5 x_7^3,\; x_4^4 - x_6 x_8^3 \,\}. \end{aligned} \tag{12.5}$$

The first eight of these ten binomials are a minimal generating set of $I_\mathcal{A}$. From the initial terms in $\mathcal{G}$ we infer that the projective variety $Y_\mathcal{A} \subset P^7$ is a threefold of degree 16. We compare the derivations of this Gröbner basis using Algorithms 4.5, 12.3 and 12.6.

- In Algorithm 4.5 we form the auxiliary ideal

$$\begin{aligned} J \quad = \quad \langle\, & x_1 - t_1 t_2^2 t_3^3 t_4^4,\; x_2 - t_1^2 t_2^3 t_3^4 t_4,\; x_3 - t_1^3 t_2^4 t_3 t_4^2,\; x_4 - t_1^4 t_2 t_3^2 t_4^3, \\ & x_5 - t_2 t_3^4 t_4^5,\; x_6 - t_1 t_2^4 t_3^5,\; x_7 - t_1^4 t_2^5 t_4,\; x_8 - t_1^5 t_3 t_4^4 \,\rangle. \end{aligned}$$

  Let $\prec$ be the purely lexicographic term order given by $t_1 > t_2 > t_3 > t_4 > x_1 > \cdots > x_8$. The reduced Gröbner basis $\mathcal{G}_\prec$ of J in this order has 207 elements, which is more than twenty times the cardinality of $\mathcal{G} = \mathcal{G}_\prec \cap k[\mathbf{x}]$. If we choose a more clever block elimination order instead, then the reduced Gröbner basis of J has 200 elements.

- In steps 1 and 2 of Algorithm 12.3 we compute a reduced lattice basis for $ker(\mathcal{A})$:

$$\mathcal{B}_{red} \quad = \quad \text{rows of} \begin{pmatrix} 0 & 1 & 0 & 1 & 0 & -1 & 0 & -1 \\ 0 & 1 & 0 & -3 & 0 & 0 & 0 & 2 \\ 1 & 0 & 1 & 0 & -1 & 0 & -1 & 0 \\ 2 & 0 & -2 & 0 & -1 & 0 & 1 & 0 \end{pmatrix}.$$

In Step 3 we form the binomial ideal associated with $\mathcal{B}_{red}$:

$$J_0 \quad = \quad \langle\, x_2x_4 - x_6x_8\,,\; x_2x_8^2 - x_4^3\,,\; x_1x_3 - x_5x_7\,,\; x_1^2x_7 - x_3^2x_5 \,\rangle.$$

Entering the loop in Step 4, we first compute the reduced Gröbner basis for J_0 with respect to reverse lexicographic order given by $x_1 < x_2 < \cdots < x_8$. The result is

$$\mathcal{G}_0 \quad = \quad \big\{\, x_6x_8 - x_2x_4,\; x_4^3 - x_2x_8^2,\; x_5x_7 - x_1x_3,\; x_3^2x_5 - x_1^2x_7,\; x_1x_3^3 - x_1^2x_7^2 \big\}.$$

Dividing the last binomial in $\mathcal{G}_0$ by x_1, we obtain five generators for $J_1 = (J_0 : x_1) = (J_0 : x_1^\infty)$. Continuing with x_2 as the smallest variable, we find that x_2 is a non-zerodivisor already, i.e., $J_2 = (J_1 : x_2) = J_1$. In the third step we compute

$$\begin{aligned} J_3 &= (J_1 : x_3) = (J_1 : x_3^\infty) \\ &= \langle\, x_2x_8^2 - x_4^3,\; x_1^2x_7 - x_3^2x_5,\; x_6x_8 - x_2x_4,\; x_5x_7 - x_1x_3, \\ &\qquad x_1x_7^2 - x_3^3,\; x_1^3 - x_3x_5^2,\; x_2^2x_4x_8 - x_4^3x_6,\; x_2^3x_4^2 - x_4^3x_6^2 \,\rangle. \end{aligned} \tag{12.6}$$

We remove the common factors x_4 and x_4^2 from the last two binomials to get a generating set for J_4. In the remaining four iterations of Step 4 we find that x_5, x_6, x_7 and x_8 are non-zerodivisors modulo J_4. This implies $J_4 = I_\mathcal{A}$. The reader may wish to compare (12.6) with the first eight binomials in (12.5). Step 5 of Algorithm 12.3 now transforms these eight binomials into the Gröbner basis (12.5).

- In Step 1 of Algorithm 12.6 we select the "inversion set" $\{i_1, \ldots, i_r\} = \{5, 6, 7, 8\}$. In Steps 2 and 3 we select a suitable lattice basis $\mathcal{B}$ for the kernel of $\mathcal{A}_{5678}$, say,

$$\mathcal{B} \quad = \quad \text{rows of} \begin{pmatrix} 1 & 1 & 1 & 1 & 1 & 1 & 1 & 1 \\ 0 & 1 & 0 & -3 & 0 & 0 & 0 & -2 \\ 1 & 0 & 1 & 0 & 1 & 0 & 1 & 0 \\ 2 & 0 & -2 & 0 & 1 & 0 & -1 & 0 \end{pmatrix},$$

and we form the associated toric ideal:

$$I_{\mathcal{A}_{5678}} \quad = \quad \langle\; x_1x_2x_3x_4x_5x_6x_7x_8 - 1\,,\; x_4^3x_8^2 - x_2\,,\; x_1x_3x_5x_7 - 1\,,\; x_1^2x_5 - x_3^2x_7 \;\rangle.$$

The loop in Step 5 has four iterations. In the first we compute the reduced Gröbner basis of $I_{\mathcal{A}_{5678}}$ for the lexicographic order given by $x_8 > x_7 > x_6 > x_5 > x_4 > x_3 > x_2 > x_1$:

$$\begin{aligned} \mathcal{G}_{5678} \quad = \quad & \big\{ x_1x_3x_5x_7 - 1, x_2x_4x_6x_8 - 1, \\ & x_1^3x_5^2 - x_3, x_2^3x_6^2 - x_4, x_3^2x_7 - x_1^2x_5, x_4^2x_8 - x_2^2x_6 \big\}. \end{aligned}$$

We switch the variable x_8 in $\mathcal{G}_{5678}$ to get generators for the next ideal:

$$I_{\mathcal{A}_{567}} = \langle x_1x_3x_5x_7 - 1, x_2x_4x_6 - x_8, \\ x_1^3x_5^2 - x_3, x_2^3x_6^2 - x_4, x_3^2x_7 - x_1^2x_5, x_4^2 - x_2^2x_6x_8\rangle.$$

By two more lexicographic Gröbner bases computations, we find the next two ideals:

$$I_{\mathcal{A}_{56}} = \langle\, x_1x_3x_5 - x_7,\ x_3^2 - x_1^2x_5x_7,\ x_2^3x_6^2 - x_4, \\ x_2x_4x_6 - x_8,\ x_2^2x_6x_8 - x_4^2,\ x_5^2x_1^3 - x_3,\ x_4^3 - x_2x_8^2\,\rangle$$

and

$$I_{\mathcal{A}_5} = \langle\, x_2^3 - x_4x_6^2,\ x_2x_4 - x_8x_6,\ x_2^2x_8 - x_4^2x_6,\ x_1^3x_5^2 - x_3, \\ x_1x_3x_5 - x_7,\ x_1^2x_5x_7 - x_3^2,\ x_4^3 - x_2x_8^2,\ x_3^3 - x_1x_7^2\,\rangle.$$

The final output ideal $I_{\mathcal{A}}$ is obtained from $I_{\mathcal{A}_5}$ by switching the variable x_5. ∎

12.B. Truncated Gröbner bases

Our next topic is truncated Gröbner bases for toric ideals. This extends the familiar notion of truncated Gröbner bases for a homogeneous ideal $J \subset k[\mathbf{x}]$. We write $J_{\leq D} := \bigoplus_{e=0}^{D} J_e$ for the (finite-dimensional) vector space of all polynomials of degree at most D in J. A subset $\mathcal{G}$ of $J_{\leq D}$ is a *D-Gröbner basis* of J with respect to a term order $\prec$ if $in_\prec(J)_{\leq D} = \langle in_\prec(\mathcal{G})\rangle_{\leq D}$. Basic properties of D-Gröbner bases can be found in Section 10.2 of (Becker & Weispfenning 1993). We assume that the reader knows how to compute them (for instance, with the command `set autodeg D` in MACAULAY).

Let $\mathcal{A} \subset \mathbf{N}^d \backslash \{0\}$ as before. Consider the $\mathbf{N}\mathcal{A}$-grading of $k[\mathbf{x}]$ via $deg(x_i) = \mathbf{a}_i$. There is a natural partial order $\leq$ on the semigroup $\mathbf{N}\mathcal{A}$, defined by $\mathbf{b}' \leq \mathbf{b}$ if and only if $\mathbf{b} - \mathbf{b}' \in \mathbf{N}\mathcal{A}$. A subset Ω of $\mathbf{N}\mathcal{A}$ is called an *order ideal* provided $\mathbf{b} \in \Omega$ and $\mathbf{b}' \leq \mathbf{b}$ implies $\mathbf{b}' \in \Omega$.

Remark 12.8. *Let Ω be an order ideal in $\mathbf{N}\mathcal{A}$. Then the k-linear span of all monomials $\mathbf{x}^\mathbf{u}$ whose degree $\pi(\mathbf{u})$ does not lie in Ω is a monomial ideal in $k[\mathbf{x}]$.*

Let J be any ideal in $k[\mathbf{x}]$ which is $\mathbf{N}\mathcal{A}$-homogeneous, such as the toric ideal $I_\mathcal{A}$, its initial ideals, or its binomial subideals of the form (12.3). We write $J_\Omega := \bigoplus_{\mathbf{b}\in\Omega} J_\mathbf{b}$ for the vector space of all polynomials in J whose degree lies in an order ideal Ω.

Proposition 12.9. *Let $\Omega, J, \prec$ as above. For a subset $\mathcal{G} \subset J_\Omega$ the following are equivalent:*
(1) The vector spaces $in_\prec(J)_\Omega$ and $\langle in_\prec(\mathcal{G})\rangle_\Omega$ coincide.
(2) Every polynomial $f \in J_\Omega$ reduces to zero with respect to $\mathcal{G}$.
(3) Every polynomial $f \in k[\mathbf{x}]_\Omega$ has a unique normal form with respect to $\mathcal{G}$.
(4) There exists a Gröbner basis $\mathcal{G}'$ of J with respect to $\prec$ such that $\mathcal{G} = \mathcal{G}' \cap k[\mathbf{x}]_\Omega$.

The proof is left to the reader. We say that $\mathcal{G}$ is an *Ω-Gröbner basis* for J with respect to $\prec$ if the four equivalent conditions of Proposition 12.9 hold. The usual definition of *reduced* Gröbner bases extends to the Ω-case. Starting from any generating set of J, we can compute the reduced Ω-Gröbner basis using the *truncated Buchberger algorithm*, that is, by ignoring all S-pairs whose degree lies outsice Ω. This is discussed in detail for $d = 1$ in (Becker & Weispfenning 1993, pages 473ff.). The general case $d > 1$ is analogous.

One new issue arising for $d > 1$ is how to test membership in Ω. For instance, the membership test for a *principal order ideal* $\Omega = \{\mathbf{b}' \in \mathbf{N}\mathcal{A} : \mathbf{b}' \leq \mathbf{b}\}$ can be as hard as solving a general integer program. In practical applications of truncated Gröbner bases it is therefore essential to select an order ideal Ω which admits an easy membership test. Here are two possibilities for good choices of order ideals:

- Choose integers $n_1, \ldots, n_d$ and set $\Omega = \{(b_1, \ldots, b_d) \in \mathbf{N}\mathcal{A} : b_i \leq n_i\}$.
- Arrange for the monomial ideal in Remark 12.8 to have few generators.

Truncated Gröbner bases for the toric ideal $I_\mathcal{A}$ are of considerable interest for integer programming (cf. Chapter 5). In this application the selection of an order ideal Ω corresponds to imposing restrictions on the possible right hand side vectors. In complete analogy to Theorem 5.5, we obtain the following corollary to Proposition 12.9.

Corollary 12.10. *Let $\mathcal{G} \subset ker(\pi)$ and $\prec$ any term order on $\mathbf{N}^n$. The directed graph $\pi^{-1}(\mathbf{b})_{\mathcal{G},\prec}$ has a unique sink for every $\mathbf{b} \in \Omega$ if and only if the set of binomials $\{\mathbf{x}^{\mathbf{v}^+} - \mathbf{x}^{\mathbf{v}^-} : \mathbf{v} \in \mathcal{G}\}$ is an Ω-Gröbner basis for the toric ideal $I_\mathcal{A}$ with respect to $\prec$.*

A direct consequence of Corollary 12.10 is an algorithm for solving integer programs with fixed matrix, fixed cost function and bounded right hand side. In the formulation of Algorithm 5.6 simply replace $\mathbf{N}\mathcal{A}$ by Ω and "Gröbner basis" by "Ω-Gröbner basis".

Example 12.7. (continued) Consider the family of integer programs defined by the matrix $\mathcal{A}$ and given lexicographic term order. Suppose we are interested only in right hand side vectors $\mathbf{b} = (b_1, b_2, b_3, b_4)$ which satisfy $b_1 \leq 8$ and $b_2 \leq 8$. This condition defines an order ideal Ω in $\mathbf{N}\mathcal{A}$. The reduced Ω-Gröbner basis equals $\{x_1^3 - x_3x_5^2,\ x_1x_3 - x_5x_7,\ x_2x_4 - x_6x_8\}$. Indeed, the degrees of these three binomials are $(3,6,9,12)$, $(4,6,4,6)$ and $(6,4,6,4)$, while the degrees of the other seven binomials in (12.5) lie outside of Ω. ∎

There are two noteworthy facts about Ω-Gröbner basis for the Lawrence matrix $\Lambda(\mathcal{A}) = \begin{pmatrix} \mathcal{A} & \mathbf{0} \\ \mathbf{1} & \mathbf{1} \end{pmatrix}$ considered in (7.1). First, a truncated version of Algorithm 7.2 computes a *universal Ω-Gröbner basis* of $I_\mathcal{A}$. By this we mean a finite subset of $I_\mathcal{A}$ which is an Ω-Gröbner basis simultaneously for all term orders. Secondly, one often encounters integer programs whose solutions $\mathbf{u}$ are required to be vectors in $\{0,1\}^n$. This condition can be coded into an order ideal Ω in the semigroup $\mathbf{N}\Lambda(\mathcal{A}) \subset \mathbf{N}^{n+d}$. Namely, take

$$\Omega \quad := \quad \{\,(b_1, \ldots, b_d, v_1, \ldots, v_n) \in \mathbf{N}\Lambda(\mathcal{A}) \;:\; 0 \leq v_1, \ldots, v_n \leq 1\} \qquad (12.7)$$

Then the universal Ω-Gröbner basis provides a test set for all integer programs of the form

$$\text{Minimize} \quad \mathbf{c} \cdot \mathbf{u} \quad \text{subject to} \quad \mathcal{A} \cdot \mathbf{u} = \mathbf{b}, \quad \mathbf{u} \in \{0,1\}^n.$$

Example 12.11. *(Primitive partition identities with distinct parts)*
Let $d = 1$, $\mathcal{A} = (1, 2, \ldots, n)$ and Ω as in (12.7). Using Observation 6.3 and Proposition 12.9, we see that a universal Ω-Gröbner basis for $I_{\Lambda(\mathcal{A})}$ is given by all primitive partition identities with distinct parts. Table 6-1 shows that there are seven of these

for $n=5$: $1+2=3$, $1+3=4$, $2+3=1+4$, $2+3=5$, $1+4=5$, $2+4=1+5$, and $3+4=2+5$.

12.C. Syzygies

We now turn our attention to minimal generators and higher syzygies of the toric ideal $I_{\mathcal{A}}$. These have a beautiful combinatorial-topological expression in terms of the fibers $\pi^{-1}(\mathbf{b})$. We continue to assume that $\mathcal{A}$ is an n-element subset of $\mathbf{N}^d$. The polynomial ring $k[\mathbf{x}]$ and its quotient $k[\mathcal{A}] = k[\mathbf{x}]/I_{\mathcal{A}}$ are graded by the semigroup $\mathbf{N}\mathcal{A}$ via $deg(x_i) = \mathbf{a}_i$. Consider a minimal free resolution of $k[\mathcal{A}]$ as a $k[\mathbf{x}]$-module:

$$0 \xrightarrow{\partial_r} k[\mathbf{x}]^{\beta_{r-1}} \xrightarrow{\partial_{r-1}} \cdots \xrightarrow{\partial_3} k[\mathbf{x}]^{\beta_2} \xrightarrow{\partial_2} k[\mathbf{x}]^{\beta_1} \xrightarrow{\partial_1} k[\mathbf{x}]^{\beta_0} \xrightarrow{\partial_0} k[\mathbf{x}] \longrightarrow 0. \tag{12.8}$$

Here $I_{\mathcal{A}} = image(\partial_0)$, so that $k[\mathcal{A}]$ appears as the cokernel of ∂_0. By Hilbert's Syzygy Theorem, we have $r \leq n$. Each of the $k[\mathbf{x}]$-module homomorphisms ∂_j is homogeneous of degree 0 with respect to the $\mathbf{N}\mathcal{A}$-grading. Each of the generators of the j-th syzygy module $k[\mathbf{x}]^{\beta_j}$ has a unique degree, which is an element in the semigroup $\mathbf{N}\mathcal{A}$. We write $\beta^j_{\mathbf{b}}$ for the number of generators having degree $\mathbf{b}$. The numbers $\beta^j_{\mathbf{b}}$ are the *multi-graded Betti numbers* of the toric ideal $I_{\mathcal{A}}$. We note that $\beta^0_{\mathbf{b}}$ equals the number of minimal generators of $I_{\mathcal{A}}$ having degree $\mathbf{b}$.

For each $\mathbf{b} \in \mathbf{N}\mathcal{A}$ we define a simplicial complex $\Delta_{\mathbf{b}}$ on the set $\{1, 2, \ldots, n\}$ as follows: A subset F of $\{1, \ldots, n\}$ is a face of $\Delta_{\mathbf{b}}$ if and only if $F \subseteq supp(\mathbf{u})$ for some $\mathbf{u} \in \pi^{-1}(\mathbf{b})$. Thus $\Delta_{\mathbf{b}}$ is the simplicial complex generated by the set system $supp(\pi^{-1}(\mathbf{b}))$. Equivalently,

$$\Delta_{\mathbf{b}} \quad = \quad \bigl\{ F \subseteq \{1, \ldots, n\} \; : \; \mathbf{b} - \sum_{i \in F} \mathbf{a}_i \in \mathbf{N}\mathcal{A} \bigr\}. \tag{12.9}$$

Theorem 12.12. *The multigraded Betti number $\beta^j_{\mathbf{b}}$ equals the rank of the j-th reduced homology group $\tilde{H}_j(\Delta_{\mathbf{b}}; k)$ of the simplicial complex $\Delta_{\mathbf{b}}$.*

Proof: We shall use basic facts from homological algebra. The desired Betti number can be expressed as follows:

$$\beta^j_{\mathbf{b}} \quad = \quad dim_k\bigl(Tor^{j+1}_{k[\mathbf{x}]}(k[\mathcal{A}] \, , \, k)_{\mathbf{b}} \bigr).$$

Here the field k is given the structure of a $k[\mathbf{x}]$-module via $k \simeq k[\mathbf{x}]/M$, where $M = \langle x_1, \ldots, x_n \rangle$. We have the following isomorphism of $\mathbf{N}\mathcal{A}$-graded $k[\mathbf{x}]$-modules

$$Tor^{j+1}_{k[\mathbf{x}]}(k[\mathcal{A}] \, , \, k) \quad \simeq \quad Tor^{j+1}_{k[\mathbf{x}]}(k \, , \, k[\mathcal{A}]).$$

The module on the right hand side is computed as follows. Form the minimal free resolution of k as a $k[\mathbf{x}]$-module, tensor it with $k[\mathcal{A}]$, and then take the j-th homology module of the resulting complex. The minimal free resolution of $k = k[\mathbf{x}]/M$ is the Koszul complex

$$0 \to \wedge_n k[\mathbf{x}]^n \to \wedge_{n-1} k[\mathbf{x}]^n \to \cdots \to \wedge_2 k[\mathbf{x}]^n \to k[\mathbf{x}]^n \to k[\mathbf{x}] \to 0.$$

Tensoring this exact sequence with the $k[\mathbf{x}]$-module $k[\mathcal{A}]$, we get the complex

$$0 \to \wedge_n k[\mathcal{A}]^n \to \wedge_{n-1} k[\mathcal{A}]^n \to \cdots \to \wedge_2 k[\mathcal{A}]^n \to k[\mathcal{A}]^n \to k[\mathcal{A}] \to 0, \tag{12.10}$$

where the j-th differential $\wedge_{j+1}k[\mathcal{A}]^n \to \wedge_j k[\mathcal{A}]^n$ is given by the formula

$$e_{i_0} \wedge e_{i_1} \wedge \cdots \wedge e_{i_j} \;\mapsto\; \sum_{s=0}^{j} (-1)^s \cdot \mathbf{t}^{\mathbf{a}_{i_s}} \cdot e_{i_0} \wedge \cdots \wedge \widehat{e_{i_s}} \wedge \cdots \wedge e_{i_j}. \tag{12.11}$$

The restriction of the Koszul complex (12.10) to its degree $\mathbf{b}$ component is the following complex of finite-dimensional k-vector spaces

$$\cdots \to \bigoplus_{i<j<l} k[\mathcal{A}]_{\mathbf{b}-\mathbf{a}_i-\mathbf{a}_j-\mathbf{a}_l} \to \bigoplus_{i<j} k[\mathcal{A}]_{\mathbf{b}-\mathbf{a}_i-\mathbf{a}_j} \to \bigoplus_{i=1}^{n} k[\mathcal{A}]_{\mathbf{b}-\mathbf{a}_i} \to k[\mathcal{A}]_{\mathbf{b}} \to 0. \tag{12.12}$$

It follows directly from the definition (12.9) that for any subset F of $\{1, 2, \ldots, n\}$,

$$k[\mathcal{A}]_{\mathbf{b}-\Sigma_{i\in F}\mathbf{a}_i} \quad = \quad \begin{cases} k & \text{if } F \text{ is a face of } \Delta_{\mathbf{b}}\,, \\ 0 & \text{otherwise}\,. \end{cases}$$

Hence we can identify the j-th term in the complex (12.12) with the k-span of the oriented j-dimensional faces of $\Delta_{\mathbf{b}}$. The restriction of the differential (12.11) to degree $\mathbf{b}$ is the usual boundary operator of simplicial topology. Therefore (12.12) is the augmented oriented chain complex of $\Delta_{\mathbf{b}}$ and its j-th reduced homology group equals $\tilde{H}_j(\Delta_{\mathbf{b}}; k)$. This completes the proof. ∎

Corollary 12.13. *The toric ideal $I_{\mathcal{A}}$ has a minimal generator in degree $\mathbf{b}$ if and only if the simplicial complex $\Delta_{\mathbf{b}}$ is disconnected.*

Example 12.7. (continued)
The computation of the minimal free resolution (12.8) in MACAULAY looks like this:

```
% <ring 8 x[1]-x[8] r
% <ideal i x[2]x[8]2-x[4]3 x[2]3-x[4]x[6]2 x[1]x[7]2-x[3]3
           x[1]3-x[3]x[5]2 x[2]x[4]-x[6]x[8] x[1]x[3]-x[5]x[7]
           x[2]2x[8]-x[4]2x[6] x[1]2x[7]-x[3]2x[5]
% res i j
1.2.3...4....5......6......7......8......9......10......
computation complete after degree 10
% betti j
total:    1 8 24 34 24 8 1
--------------------------------------------------
      0:  1 -  -  -  -  - -
      1:  - 2  -  -  -  - -
      2:  - 6  9  2  -  - -
      3:  - -  6  8  2  - -
      4:  - -  9 24 22  8 1
```

This table gives the Betti numbers of $k[\mathcal{A}]$ in the usual total degree grading. We write $\beta_d^j := \sum_{|\mathbf{b}|=10d} \beta_{\mathbf{b}}^j$. In the second column we find $\beta_2^0 = 2$ and $\beta_3^0 = 6$. This means that $I_{\mathcal{A}}$ has eight minimal generators, 2 quadrics and 6 cubics. The higher Betti numbers are

$$\beta^1_4 = 9,\ \beta^1_5 = 6,\ \beta^1_6 = 9,\ \ \beta^2_5 = 2,\ \beta^2_6 = 8,\ \beta^2_7 = 24,$$
$$\beta^3_7 = 2,\ \beta^3_8 = 22,\ \ \beta^4_9 = 8,\ \beta^5_{10} = 1.$$

We now illustrate Theorem 12.12 by deriving the highest non-vanishing Betti number β^5_{10}. More precisely, we shall prove that

$$\beta^5_{\mathbf{b}} \quad = \quad dim_k\, \tilde{H}_5(\Delta_{\mathbf{b}}; k) \quad = \quad 1 \qquad\qquad \text{for} \quad \mathbf{b} = (25, 25, 25, 25).$$

Using Algorithm 5.7, we easily enumerate the fiber of that multi-degree:

$$\begin{aligned}\pi^{-1}(\mathbf{b}) \quad = \quad \big\{ &(00222211),\ (01232110),\ (02202112),\ (03212011),\\ &(10321201),\ (11331100),\ (12301102),\ (13311001),\\ &(20021221),\ (21031120),\ (22001122),\ (23011021),\\ &(30120211),\ (31130110),\ (32100112),\ (33110011)\big\}.\end{aligned}$$

The simplicial complex $\Delta_{\mathbf{b}} = \{345678, 234567, 235678\ldots, 123478\}$ is a triangulation of the 5-dimensional sphere. In fact, $\Delta_{\mathbf{b}}$ equals the boundary complex of the *cyclic 6-polytope with 8 vertices*; see (Ziegler 1995, Theorem 0.7) where this polytope is denoted $C_6(8)$. To identify $\Delta_{\mathbf{b}}$ with $\partial C_6(8)$, either check that the minimal non-faces of $\Delta_{\mathbf{b}}$ are precisely $\{1, 3, 5, 7\}$ and $\{2, 4, 6, 8\}$, or verify *Gale's evenness condition* for the supports of the 16 vectors in $\pi^{-1}(\mathbf{b})$. Since $\Delta_{\mathbf{b}}$ is a 5-sphere, the group $\tilde{H}_5(\Delta_{\mathbf{b}}; k)$ is k^1. ■

12.D. Localization in integer programming

Linear programming is generally much easier than integer programming. Indeed, linear programs can be solved in polynomial time, whereas there is very little hope (unless $P = NP$) that general integer programs can be solved in polynomial time. Practical experience confirms these complexity results. Therefore, when analyzing the families of integer programs in Section 5, it makes sense to assume that the solution to the linear relaxation (8.4) is already known. In fact, taking the conjunction over all right hand sides $\mathbf{b}$, we may even assume that the regular triangulation $\Delta_{\prec}$ is given to us explicitly, along with the matrix $\mathcal{A}$ and the term order $\prec = \prec_\omega$. We shall discuss algebraic techniques for speeding up Algorithm 5.6 which are based on this assumption.

We write $M := in_{\prec}(I_{\mathcal{A}})$ for the corresponding initial ideal. It is spanned by all (monomials $\mathbf{x}^{\mathbf{u}}$ representing) non-optimal points $\mathbf{u} \in \mathbf{N}^n$ with respect to $\prec$. We are interested in the *primary decomposition* of M. We assume that the reader knows how to compute the *associated primes* of a monomial ideal; see e.g. (Eisenbud 1995. p. 111). By Theorem 8.3 and Corollary 8.4, the *minimal primes* of M are in bijection with the maximal faces of $\Delta_{\prec}$, while the *embedded primes* correspond to certain lower-dimensional faces of $\Delta_{\prec}$.

Let $\mathbf{x}^{\mathbf{u}} \notin M$ be a standard monomial. A subset X of the variable set $\{x_1, \ldots, x_n\}$ (or, equivalently, a subset of $\mathcal{A}$) is said to be *compatible with* $\mathbf{x}^{\mathbf{u}}$ if every monomial in the vector space $\mathbf{x}^{\mathbf{u}} \cdot k[X]$ is standard. If X is compatible with $\mathbf{x}^{\mathbf{u}}$ and has maximal cardinality with this property, then we say that X is *associated* with $\mathbf{x}^{\mathbf{u}}$.

Lemma 12.14.
(a) *If X is compatible with a standard monomial $\mathbf{x}^u$, then the set X is a simplex in $\Delta_\prec$.*
(b) *If X is associated with a standard monomial $\mathbf{x}^u$, then* $P_X := \langle \{x_1, \ldots, x_n\} \backslash X \rangle$ *is an associated prime ideal of M.*

Proof: The compatibility condition is equivalent to

$$\mathbf{x}^{\mathbf{u}} \cdot k[X] \cap M = \{0\} \qquad \Longleftrightarrow \qquad (M : \mathbf{x}^{\mathbf{u}}) \subseteq P_X. \tag{12.13}$$

If this holds, then there exists a superset of variables $X' \supseteq X$ such that $P_{X'}$ is a minimal prime of $(M : \mathbf{x}^{\mathbf{u}})$. This implies that $P_{X'}$ is an associated prime of M. Hence X' is a simplex in $\Delta_\prec$, and so is its subset X. If X is associated with $\mathbf{x}^{\mathbf{u}}$, then X is maximal with the property (12.13). In this case $X = X'$ and hence P_X is an associated prime of M. ∎

Our principal goal is to solve the integer programming problem:

$$\textit{Find the } \prec\textit{-minimal vector } \quad \mathbf{u} \in \mathbf{N}^n \quad \text{subject to} \quad \mathcal{A} \cdot \mathbf{u} = \mathbf{b}. \tag{12.14}$$

A subset of variables X is *compatible* (resp. *associated*) with the right hand side $\mathbf{b}$ if X is compatible (resp. associated) with $\mathbf{x}^{\mathbf{u}}$, where $\mathbf{u}$ is the optimal solution to (12.14). We abbreviate $X^c := \{x_1, \ldots, x_n\} \backslash X$ and consider the localized polynomial ring $k(X)[X^c]$. By restricting to monomials in X^c alone, $\prec$ defines a term order on $k(X)[X^c]$. Let $\mathcal{G}_X$ be the reduced Gröbner basis with respect to $\prec$ of the image of the toric ideal $I_\mathcal{A}$ in $k(X)[X^c]$.

Theorem 12.15. *Let $\mathbf{u}'$ be any vector in $\mathbf{N}^n$ satisfying $\mathcal{A} \cdot \mathbf{u}' = \mathbf{b}$, and let X be any simplex in the triangulation $\Delta_\prec$. Then the following three statements are equivalent:*
(a) *The simplex X is compatible to $\mathbf{b}$.*
(b) *The normal form of $\mathbf{x}^{\mathbf{u}'}$ with respect to $\mathcal{G}_X$ is a monomial in $k[x_1, \ldots, x_n]$.*
(c) *The normal form of $\mathbf{x}^{\mathbf{u}'}$ with respect to $\mathcal{G}_X$ is the optimal solution to (12.14).*

Proof: Let $\mathbf{u}$ be the optimal solution of (12.14), and let ΠX denote the product of all variables in X. The localized initial ideal equals

$$\langle\, in_\prec(\mathcal{G}_X) \rangle \quad = \quad (\, M \,:\, (\Pi X)^\infty\,). \tag{12.15}$$

Statement (c) holds if and only if $\mathbf{x}^{\mathbf{u}}$ does not lie in (12.15) if and only if (12.13) holds if and only if (a) holds. Clearly (c) implies (b). Suppose that (b) holds, and let $\mathbf{x}^{\mathbf{v}}$ be the normal form of $\mathbf{x}^{\mathbf{u}'}$ with respect to $\mathcal{G}_X$. Then $\mathbf{x}^{\mathbf{v}} \preceq \mathbf{x}^{\mathbf{u}} \preceq \mathbf{x}^{\mathbf{u}'}$. Since $\mathbf{x}^{\mathbf{v}} \in k[x_1, \ldots, x_n]$, we have $\mathbf{x}^{\mathbf{u}} - \mathbf{x}^{\mathbf{v}} \in I_\mathcal{A}$. These two facts imply $\mathbf{x}^{\mathbf{u}} = \mathbf{x}^{\mathbf{v}}$, which means that (c) holds. ∎

The utility of this theorem lies in the fact that $\mathcal{G}_X$ is often easier to compute than $\mathcal{G} = \mathcal{G}_\emptyset$, the Gröbner basis of $I_\mathcal{A}$ in $k[x_1, \ldots, x_n]$. Computing $\mathcal{G}_X$ is most efficient when X is a maximal simplex $\Delta_\prec$, because in this case $\langle \mathcal{G}_X \rangle = I_\mathcal{A} \cdot k(X)[X^c]$ is a zero-dimensional ideal. This heuristic shortcut to solving (12.14) applies to almost all right hand sides $\mathbf{b}$.

Proposition 12.16. *Suppose that $I_\mathcal{A}$ is positively graded. As the degree increases, the fraction of right hand sides* **b** *which are compatible to maximal simplices of $\Delta_\prec$ tends to* 1.

Proof: We identity the right hand sides $\mathbf{b} \in \mathbf{N}\mathcal{A}$ with the standard monomials $\mathbf{x}^\mathbf{u} \not\in M$. Their number in degree r equals the value of the Hilbert function $H_M(r)$. Let T denote the intersection of all *top-dimensional* primary components of M (i.e., primary components whose radical is a minimal prime of M). We have $M \subseteq T$. A monomial $\mathbf{x}^\mathbf{u}$ lies in $T \backslash M$ if and only if no minimal prime of M is associated to $(M : \mathbf{x}^\mathbf{u})$. This means that no maximal simplex of $\Delta_\prec$ is compatible with $\mathbf{x}^\mathbf{u}$. We conclude that the fraction of right hand sides **b** compatible to maximal simplices in degree r equals $H_T(r)/H_M(r)$. Since M and T are ideals of the same dimension and degree, we have $lim_{r\to\infty} H_T(r)/H_M(r) = 1$. ∎

Example 12.7. (continued)
The initial terms in (12.5) generate the monomial ideal

$$\begin{aligned} M &= (x_1^3, x_1^2x_7, x_1x_3, x_1x_7^2, x_2^3, x_2^2x_8, x_2x_4, x_2x_8^2, x_3^4, x_4^4) \\ &= Q_1 \cap Q_2 \cap Q_3 \cap Q_4, \end{aligned}$$

whose primary decomposition consists of the four ideals

$$\begin{aligned} Q_1 &= \langle x_1, x_2, x_3^4, x_4^4 \rangle, & Q_2 &= \langle x_1^3, x_1^2x_7, x_2, x_3, x_4^4, x_7^2 \rangle, \\ Q_3 &= \langle x_1, x_2^3, x_2^2x_8, x_3^4, x_4, x_8^2 \rangle, \quad \text{and} & Q_4 &= \langle x_1^3, x_1^2x_7, x_2^3, x_2^2x_8, x_3, x_4, x_7^2, x_8^2 \rangle. \end{aligned}$$

The associated primes $P_i = Rad(Q_i)$ are

$$P_1 = \langle x_1, x_2, x_3, x_4 \rangle, \; P_2 = P_1 + \langle x_7 \rangle, \; P_3 = P_1 + \langle x_8 \rangle, \text{ and } P_4 = P_1 + \langle x_7, x_8 \rangle.$$

There is only one minimal prime $P_1 = Rad(M)$. Hence $T = Q_1$ and $\Delta_\prec = \{\{5,6,7,8\}\}$. The fact that the triangulation $\Delta_\prec$ has only one maximal simplex means that the linear program (8.4) is trivial to solve: for each $\mathbf{b} \in pos(\mathcal{A})$, the LP-optimum is the unique vector $\mathbf{v} = (0,0,0,0,v_5,v_6,v_7,v_8) \in \mathbf{R}^8$ satisfying $\mathcal{A}\mathbf{v} = \mathbf{b}$.

Suppose we wish to solve the integer program (12.14) for a given feasible solution but the Gröbner basis $\mathcal{G}$ in (12.5) is not yet known or too hard to compute. The heuristic suggested by Proposition 12.16 is to use instead the local Gröbner basis $\mathcal{G}_X$ for $X = \{x_5, x_6, x_7, x_8\}$. Applying a variant of Algorithm 12.3 in $k(X)[x_1, x_2, x_3, x_4]$, we find

$$\mathcal{G}_X = \{ x_1 - x_3^3 x_7^{-2}, \; x_2 - x_4^3 x_8^{-2}, \; x_3^4 - x_5 x_7^3, \; x_4^4 - x_6 x_8^3 \}.$$

Note that $T = Q_1 = \langle in_\prec(\mathcal{G}_X) \rangle = (M : (x_5x_6x_7x_8)^\infty)$.

We consider two given feasible solutions of degree $r = 1000$:

$$\mathbf{v}^{(1)} = (2,76,1,11,231,372,1,206) \quad \text{and} \quad \mathbf{v}^{(2)} = (2,76,1,11,231,372,0,207).$$

The normal form of $\mathbf{x}^{\mathbf{v}^{(i)}}$ with respect to $\mathcal{G}_X$ is found to be $\mathbf{x}^{\mathbf{u}^{(i)}}$ where

$$\mathbf{u}^{(1)} = (0,0,3,3,232,431,0,231) \quad \text{and} \quad \mathbf{u}^{(2)} = (0,0,3,3,232,431,-1,232).$$

In the first case our heuristic succeeded in solving (12.14): the monomial $\mathbf{x}^{\mathbf{u}^{(1)}}$ is the normal form of $\mathbf{x}^{\mathbf{v}^{(1)}}$ with respect to the "unknown" Gröbner basis $\mathcal{G}$, by Theorem

12.15. In the second case it was unsuccessful. The rate of success of this heuristic equals

$$\frac{H_T(r)}{H_M(r)} \quad = \quad \frac{8/3r^3 - 8r^2 + 76/3r - 20}{8/3r^3 + 4r^2 - 17/3r + 7} \qquad \text{in degree } r \geq 3$$

The probability of hitting a bad fiber in degree 1000 is $1 - \frac{H_T(1000)}{H_M(1000)} = 0.00448\cdots$.
■

Returning to our general discussion, consider the unlucky event that the localization heuristic fails. Then the negative exponents in the normal form $x_1^{j_1}x_2^{j_2}\cdots x_n^{j_n}$ modulo $\mathcal{G}_X$ suggest the following next step: Replace X by the subsimplex $X\backslash\{x_\nu : j_\nu < 0\}$ and redo the local Gröbner basis computation and reduction. For instance, replacing $X = \{5,6,7,8\}$ by $X = \{5,6,8\}$ for $\mathbf{v}^{(2)}$ in the above example yields the correct solution $(1,0,0,3,232,431,1,232)$. The trade off is that $\mathcal{G}_{568}$ has cardinality seven.

The efficiency of the heuristic is improved by working in $k[X^c]$ instead of $K(X)[X^c]$. Indeed, consider the map $\rho : k(X)[X^c] \to k[X^c]$ which sends each variable $x_i \in X$ to the constant 1. Since X is a simplex in $\Delta_\prec$, it follows that the restriction of ρ to the set of irreducible binomials in $I_\mathcal{A} \cdot k(X)[X^c]$ is injective. Moreover, the inverse map is easy to compute by homogenization. This means in practise that one computes the Gröbner basis $\mathcal{G}'_X := \rho(\mathcal{G}_X)$ instead of $\mathcal{G}_X$.

Exercises:

(1) Show that the Gröbner basis $\mathcal{G}'$ in Lemma 12.1 need not be reduced, even if $\mathcal{G}$ is reduced.

(2) Prove Proposition 12.9.

(3) Consider the following vertex set of a planar pentagon,

$$\mathcal{A} \quad = \quad \begin{pmatrix} 0 & 6 & 11 & 4 & 2 \\ 0 & 1 & 5 & 9 & 5 \\ 1 & 1 & 1 & 1 & 1 \end{pmatrix},$$

and let $\prec$ be the lexicographic order given by $x_1 \prec x_2 \prec x_3 \prec x_4 \prec x_5$.
  (a) Compute the reduced Gröbner basis for $I_\mathcal{A}$ using both Algorithms 12.3 and 12.6.
  (b) Compute a primary decomposition of $in_\prec(I_\mathcal{A})$.
  (c) The maximal faces of the regular triangulation $\Delta_\prec$ are the three triangles 123, 134 and 145. Compute the corresponding local Gröbner bases $\mathcal{G}_{123}, \mathcal{G}_{234}$ and $\mathcal{G}_{345}$.
  (d) Find a right hand side vector $\mathbf{b} = (b_1, b_2, b_3)$ with $b_3 = 1000$ such that the local heuristic fails simultaneously for all three triangles in $\Delta_\prec$.
  (e) What is the probability for the unlucky event in (d) to happen if $\mathbf{b}$ is chosen at random from the uniform distribution on $1000 \cdot \mathcal{A}$?

(4) Compute the minimal free resolution (12.8) of $I_\mathcal{A}$ for the matrix $\mathcal{A}$ in the previous exercise. Verify Theorem 12.12 for the minimal generators and the second syzygies.

(5) Characterize integer programs with the property that $in_\prec(I_\mathcal{A})$ has no embedded components.

(6) Explain how truncated Gröbner bases can be used to solve integer programming problems of the form

$$\text{Minimize} \quad \mathbf{c} \cdot \mathbf{u} \quad \text{subject to} \quad \mathcal{A} \cdot \mathbf{u} = \mathbf{b}, \quad \mathbf{u} \in \{0, 1, \ldots, r\}^n.$$

Notes:
Algorithm 12.6 is due to Di Biase and Urbanke (1995). Both Algorithm 12.3 and 12.6 are implemented in Hosten's program GRIN ("**Gr**öbner bases for **In**teger Programming"). An experimental discussion and comparison to existing integer programming software (CPLEX) is given in (Hosten & Sturmfels 1995). GRIN is available via anonymous ftp from `ftp.orie.cornell.edu`. Truncated Gröbner bases for integer programming have been introduced by Thomas and Weismantel (1995). In this work the combinatorial algorithms of (Urbaniak, Weismantel & Ziegler 1994) are extended and placed in an algebraic setting. The characterization of Koszul homology of toric ideals (Theorem 12.12) appears for $d = 1$ in (Campillo & Marijuan 1991) and for general d in (Campillo & Pison 1993). The passage to the local Gröbner bases $\mathcal{G}_X$ in Theorem 12.15 is equivalent to the relaxation to the *group problem in integer programming* as presented on page 364 in (Schrijver 1986). This equivalence was suggested in Section 6 of (Sturmfels, Weismantel & Ziegler 1995).

CHAPTER 13

Toric Varieties in Algebraic Geometry

The theory of toric varieties plays an important role at the crossroads of geometry, algebra and combinatorics. It provides a fertile testing ground for general theories in algebraic geometry (and in symplectic geometry, and in topology, and ...). It is the objective of this chapter to establish a connection to toric varieties as they are defined and used by algebraic geometers. We shall assume that the reader is familiar with the books of Fulton (1993) and Oda (1988). Notation and results from these books will be used.

Starting with Chapter 4 we have chosen the name "toric variety" for the zero set of any toric ideal $I_{\mathcal{A}}$. In other words, we defined a *toric variety* to be an affine or projective variety which is parametrized by a set $\mathcal{A}$ of monomials. This nomenclature disagrees with the standard definition used in algebraic geometry, which goes as follows: a *toric variety* is a *normal* variety X that contains an algebraic torus $T \simeq (k^*)^d$ as a dense open subset, together with an action $T \times X \to X$ of T on X that extends the natural action of T on itself. (In this chapter we shall make the simplifying assumption that k is an algebraically closed field of characteristic zero.) The crucial requirement here is that X is *normal*. In the following we shall prepend this adjective for the sake of utmost clarity.

Let X be a normal toric variety which is either affine or projective, and suppose further that X is endowed with an explicit embedding into affine or projective space. Then X is defined by a toric ideal $I_{\mathcal{A}}$, where $\mathcal{A}$ belongs to a special class of vector configurations. Our first goal is to identify these configurations.

We fix a lattice $N \simeq \mathbf{Z}^d$ and its dual lattice $M = N^\vee$. Let σ denote a strongly convex (rational polyhedral) cone in the corresponding vector space $N_{\mathbf{Q}} \simeq \mathbf{Q}^d$. Its polar dual

$$\sigma^\vee \quad := \quad \{\, \mathbf{u} \in M_{\mathbf{Q}} \;:\; \mathbf{u} \cdot \mathbf{v} \geq 0 \text{ for all } \mathbf{v} \in \sigma \,\}.$$

is a d-dimensional cone in the dual vector space. The associated semigroup

$$S_\sigma \quad := \quad \sigma^\vee \cap M \quad = \quad \{\, \mathbf{u} \in M \;:\; \mathbf{u} \cdot \mathbf{v} \geq 0 \text{ for all } \mathbf{v} \in \sigma \,\}$$

is finitely generated, and hence so is the semigroup algebra $k[S_\sigma]$. Its spectrum $X_\sigma := \mathit{Spec}\,(k[S_\sigma])$ is a normal affine toric variety. Conversely, every normal affine toric variety is isomorphic to X_σ for some cone σ; see (Oda 1988, Theorem 1.5).

Lemma 13.1. *The cone σ is d-dimensional if and only if its dual $\sigma^\vee$ is strongly convex. In this case the semigroup S_σ has a unique minimal finite generating set $\mathcal{A} \subset M \simeq \mathbf{Z}^d$.*

Proof: The cone $\sigma^\vee$ is strongly convex if and only if its lineality space $\sigma^\vee \cap (-\sigma^\vee) = \sigma^\perp$ equals $\{0\}$ if and only if σ is d-dimensional. The hypothesis $S_\sigma \cap (-S_\sigma) = \{0\}$

implies that we can define a partial order on $S_\sigma \backslash \{0\}$ by $\mathbf{u} \leq \mathbf{v} :\iff \mathbf{v} - \mathbf{u} \in S_\sigma$. Let $\mathcal{A}$ be the set of minimal elements in this partial order. Clearly, the set $\mathcal{A}$ minimally generates S_σ, and it is unique with this property. To show that $\mathcal{A}$ is finite, we need the fact that $\sigma^\vee$ is a rational cone. Let $\mathbf{a}_1, \ldots, \mathbf{a}_r$ be the first non-zero lattice points on the extreme rays of $\sigma^\vee$. Then $\mathcal{A}$ is contained in the zonotope (sum of line segments) $\sum_{i=1}^r [0, \mathbf{a}_i]$. This sum being a bounded set, it contains only finitely many lattice points. ∎

The unique minimal generating set $\mathcal{A}$ of a strongly convex semigroup S_σ is called the *Hilbert basis*. Here is an algorithm using Gröbner bases for computing Hilbert bases.

Algorithm 13.2. *(Computing the Hilbert basis for an affine toric variety)*
Input: A spanning set for a d-dimensional convex polyhedral cone σ in $N \simeq \mathbf{Z}^d$.
Output: The Hilbert basis $\mathcal{A}$ of the semigroup $S_\sigma \subset M = N^\vee$.

0. Identify both N and M with $\mathbf{Z}^d$.
1. Replace the given generators of the cone σ by a new generating set $\{\mathbf{v}_1, \ldots, \mathbf{v}_d, \mathbf{v}_{d+1}, \ldots, \mathbf{v}_m\}$ consisting only of lattice points and such that $\{\mathbf{v}_1, \ldots, \mathbf{v}_d\}$ is a lattice basis of $\mathbf{Z}^d$. Let V denote the $m \times d$-matrix whose rows are the vectors $\mathbf{v}_1, \ldots, \mathbf{v}_m$.
2. The image of V in $\mathbf{Z}^m$ is a *saturated* sublattice, i.e., $\mathbf{Z}^m / im_{\mathbf{Z}}(V)$ is free abelian. Compute an $(m - d) \times m$-integer matrix B whose kernel equals $im_{\mathbf{Z}}(V)$.
3. Compute the Graver basis $Gr_B \subset \mathbf{Z}^m$ of the matrix B using Algorithm 7.2.
4. For each <u>non-negative</u> vector $\mathbf{s} = (s_1, \ldots, s_m)$ in the Graver basis Gr_B determine and output the unique vector $\mathbf{u} \in \mathbf{Z}^d$ such that $\mathbf{u} \cdot \mathbf{v}_i = s_i$ for $i = 1, \ldots, d$.

Discussion and proof of correctness: Since σ is a rational cone, it is generated by lattice vectors, and since σ is d-dimensional, we can augment any such generating set by a lattice basis. The computations in steps 1 and 2 involve standard integer linear algebra. They can be performed using the Hermite normal form algorithm (see e.g. (Schrijver 1986)).

We replace the semigroup $S_\sigma = \{\mathbf{u} \in \mathbf{Z}^d : \mathbf{u} \cdot \mathbf{v}_i \geq 0 \text{ for } i = 1, \ldots, m\}$ by its image under the monomorphism V, which equals

$$\mathbf{N}^m \cap Im_{\mathbf{Z}}(V) \quad = \quad \mathbf{N}^m \cap ker_{\mathbf{Z}}(B). \tag{13.1}$$

The Hilbert basis of the semigroup (13.1) consists of those elements which are minimal in the componentwise partial order on $\mathbf{N}^m$. These are precisely the non-negative vectors in the Graver basis Gr_B of B. Here the Graver basis is thought of as a set of vectors in $\mathbf{Z}^m$ rather than binomials in $k[\mathbf{x}]$. The vector $\mathbf{u}$ computed in Step 4 is the unique preimage under V of $\mathbf{s}$ in S_σ. These vectors constitute the Hilbert basis of S_σ. ∎

Example 13.3. Let σ be the cone spanned by $(3,1)$ and $(1,2)$ in the plane $\mathbf{Q}^2$. We shall compute the Hilbert basis $\mathcal{A}$ of S_σ using Algorithm 13.2. Since the two given generators do not span the lattice $\mathbf{Z}^2$, in Step 1 we throw in the vector $(1,1) = \frac{1}{5} \cdot (3,1) + \frac{2}{5} \cdot (1,2) \in \sigma$. Now $\{(1,2),(1,1)\}$ is a basis of $\mathbf{Z}^2$. The two matrices constructed in Steps 1 and 2 are

$$V \quad = \quad \begin{pmatrix} 1 & 2 \\ 1 & 1 \\ 3 & 1 \end{pmatrix} \qquad \text{and} \qquad B \quad = \quad \begin{pmatrix} 2 & -5 & 1 \end{pmatrix}.$$

In Step 3 we compute the Graver basis of I_B. In vector notation this Graver basis equals

$$Gr_B \quad = \quad \{\, (5,2,0)\,,\ (2,1,1)\,,\ (1,1,3)\,,\ (0,1,5)\,,\ (1,0,-2)\,,\ (3,1,-1)\,\}.$$

In Step 4 we compute the preimages under V of the four non-negative vectors in Gr_B. This is easily done by inverting (over $\mathbf{Z}$) the 2×2-submatrix consisting of the first two rows of V. The set of four preimages equals the desired Hilbert basis:

$$\mathcal{A} \quad = \quad \{\,(-1,3)\,,\ (0,1)\,,\ (1,0)\,,\ (2,-1)\,\}.$$

The normal toric variety X_σ is the spectrum of the toric algebra $k[S_\sigma] = k[\mathcal{A}] = k[\mathbf{x}]/I_\mathcal{A}$. We conclude that X_σ is embedded in affine 4-space as the zero set of the toric ideal

$$I_\mathcal{A} \quad = \quad \langle\, x_2x_4 - x_3^2\,,\ x_1x_4 - x_2^2x_3\,,\ x_1x_3 - x_2^3\,\rangle. \qquad (13.2)$$

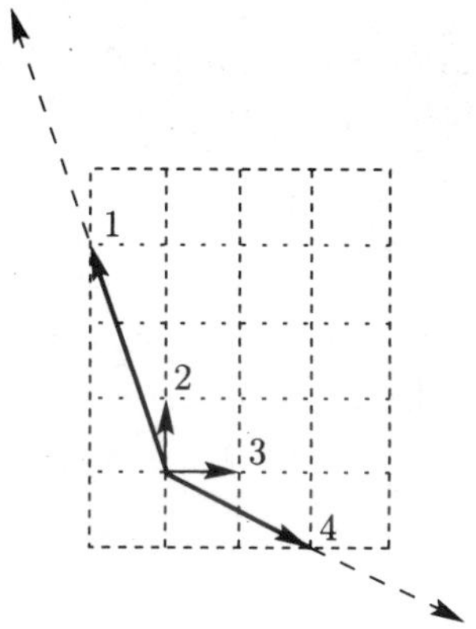

Figure 13-1. The fan in Example 13.3.

In the previous chapters we wrote $X_\mathcal{A}$ for the zero set in k^n of any toric ideal $I_\mathcal{A}$, and we called $X_\mathcal{A}$ an *affine toric variety.* This terminology is well-justified since $X_\mathcal{A}$ contains an algebraic torus T as a dense open subset, together with an action $T \times X_\mathcal{A} \to X_\mathcal{A}$ of T on $X_\mathcal{A}$ that extends the natural action of T on itself. All that is lacking is the extra requirement of *normality.* Here is an easy way to visualize the dense torus in $X_\mathcal{A}$.

Lemma 13.4. *Suppose $dim(\mathcal{A}) = d$. Then the set $X_\mathcal{A} \cap (k^*)^n$ is an algebraic group under coordinatewise multiplication isomorphic to the d-dimensional torus $T = (k^*)^d$.*

Proof: For any n-element subset $\mathcal{A}$ of $\mathbf{Z}^d$, the map

$$(k^*)^d \;\to\; X_\mathcal{A} \cap (k^*)^n, \quad \mathbf{t} = (t_1,\ldots,t_d) \;\mapsto\; (\mathbf{t}^{\mathbf{a}_1}, \mathbf{t}^{\mathbf{a}_2}, \ldots, \mathbf{t}^{\mathbf{a}_n})$$

is an epimorphism of algebraic groups. The additional requirement $dim(\mathcal{A}) = d$ guarantees that this map has a regular inverse. ∎

The issue of normality is summarized in our next proposition.

Proposition 13.5. *For a finite subset $\mathcal{A}$ of $\mathbf{Z}^d$ the following are equivalent:*
(1) The affine toric variety $X_\mathcal{A}$ is normal.
(2) The affine toric variety $X_\mathcal{A}$ is isomorphic to X_σ for some rational cone σ in $\mathbf{Q}^d$.
(3) The integral domain $k[\mathcal{A}] = k[\mathbf{x}]/I_\mathcal{A}$ is integrally closed (in its field of fractions).
(4) The semigroup $\mathbf{N}\mathcal{A}$ is normal, i.e., $\mathbf{N}\mathcal{A} = \mathbf{Z}\mathcal{A} \cap pos(\mathcal{A})$.

Proof: The conditions (1) and (3) are equivalent by definition of "normality". The implications "(2) $\Rightarrow$ (4)" and "(3) $\Rightarrow$ (4)" are obvious. Suppose that (4) holds. To prove (2) we choose an isomorphism between $\mathbf{Z}\mathcal{A}$ and $\mathbf{Z}^d$ and we replace $\mathcal{A}$ be its image under this isomorphism. Let $\tau = pos(\mathcal{A})$ be the cone generated by this set in $\mathbf{Q}^d$. Its polar dual $\sigma = \tau^\vee$ is a rational cone in $\mathbf{Q}^d$ as well. We have $\sigma^\vee = \tau^{\vee\vee} = \tau$ and $S_\sigma = \mathbf{Z}^d \cap \tau = \mathbf{Z}\mathcal{A} \cap pos(\mathcal{A}) = \mathbf{N}\mathcal{A}$. Therefore X_σ is isomorphic to $X_\mathcal{A}$. For the remaining implication "(4) $\Rightarrow$ (3)" we refer to the proposition on pages 29-30 of (Fulton 1993). ∎

Corollary 13.6. *For any finite subset $\mathcal{A}$ of $\mathbf{Z}^d$, the normalization of the affine toric variety $X_\mathcal{A}$ is the normal toric variety X_σ where $M = \mathbf{Z}\mathcal{A}$, $N = M^\vee$, and $\sigma = (pos(\mathcal{A}))^\vee$.*

Let σ and $\mathcal{A}$ be as in Lemma 13.1. Then $X_\sigma = X_\mathcal{A}$ is smooth if and only if $\mathcal{A}$ is a lattice basis of $\mathbf{Z}^d$ if and only if $I_\mathcal{A} = \{0\}$. In all other cases the given normal affine toric variety has a singularity at the origin 0. An important invariant of a singular point on a variety X is its *tangent cone.* According to (Shafarevich 1977, Section II.1.5), the tangent cone is defined by the initial forms of lowest total degree of all defining equations of X. We paraphrase this for the toric case using initial ideals.

Remark 13.7. *Let $X_\sigma = X_\mathcal{A}$ be a normal affine toric variety as in Lemma 13.1. The defining ideal of the tangent cone at the origin $0 \in X_\mathcal{A}$ is $in_{-\mathbf{e}}(I_\mathcal{A})$, where $\mathbf{e} = (1, 1, \ldots, 1)$.*

Note that $S_\sigma = \mathbf{N}\mathcal{A}$ is a strongly convex semigroup, so that the Gröbner region of $I_\mathcal{A}$ equals all of $\mathbf{R}^n$. Hence the negative vector $-\mathbf{e}$ does determine a partial term order for $I_\mathcal{A}$. Using the identifications of Theorems 8.3 and 10.10, the underlying reduced scheme of the tangent cone equals the coherent polyhedral subdivision $\Delta_{-\mathbf{e}}$ of $\mathcal{A}$. For instance, consider Example 13.3: the three binomials in (13.2) are the reduced Gröbner basis for any term order refining $-\mathbf{e} = (-1, -1, -1, -1)$. The tangent cone of this 2-dimensional toric singularity happens to be a reduced scheme. It is defined by the Stanley ideal

$$in_{(-1,-1,-1,-1)}(I_\mathcal{A}) = \langle\, x_1x_3,\, x_1x_4,\, x_3^2 - x_2x_4 \,\rangle = \langle\, x_1,\, x_3^2 - x_2x_4 \,\rangle \,\cap\, \langle\, x_3,\, x_4 \,\rangle.$$

The corresponding polyhedral subdivision of $\mathcal{A}$ is

$$\Delta_{(-1,-1,-1,-1)} = \big\{\{2,3,4\},\{1,2\}\big\}.$$

We now turn our attention to normal projective toric varieties. Following (Fulton 1993) and (Oda 1988) we fix a complete fan Δ in $N \simeq \mathbf{Z}^d$ which is the normal fan of a polytope in $M_{\mathbf{Q}}$. Such fans are called *strongly polytopal.* They are discussed in (Fulton 1993, Section 1.5) and in (Oda 1988, Section 2.4). The normal projective toric variety X_Δ is glued from the normal affine toric varieties X_σ, where σ runs over all cones in Δ. This is an abstract construction. A priori there is no concrete model of X_Δ in any projective space. To map X_Δ into projective space we must choose an ample divisor D. Let P_D be the associated polytope in $M_{\mathbf{Q}} \simeq \mathbf{Q}^d$ (Fulton 1993, Section 3.4). Then we get a map ψ_D from X_Δ onto the projective variety $Y_{\mathcal{A}}$ which is the zero set of the homogeneous toric ideal $I_{\mathcal{A}}$, where

$$\mathcal{A} \quad := \quad (P_D \cap M) \times \{1\} \quad \subset \quad M \oplus \mathbf{Z} \ \simeq \ \mathbf{Z}^{d+1}. \tag{13.3}$$

The extra coordinate "1" is essential here: in particular, it guarantees the homogeneity condition in Lemma 4.14. Our assertion that $Y_{\mathcal{A}}$ equals the image of X_Δ under the map ψ_D defined by the ample divisor D is the content of (Fulton 1993, Lemma on page 66).

In order to determine whether D is very ample, that is, whether $Y_{\mathcal{A}}$ is actually isomorphic to X_Δ, we consider the affine cover $\{X_\sigma\}_{\sigma\in\Delta}$ of X_Δ. We identify P_D with the convex hull of $\mathcal{A}$. Each d-dimensional cone $\sigma \in \Delta$ is the normal fan of the polytope P_D at a vertex $\mathbf{a}_\sigma$. In the notation of Chapter 1 this is expressed as $\sigma = \mathcal{N}_{P_D}(\{\mathbf{a}_\sigma\})$.

Lemma 13.8. *The divisor D is very ample if and only if the semigroup S_σ is generated by $\mathcal{A} - \mathbf{a}_\sigma$ for every d-dimensional cone $\sigma \in \Delta$,*

Proof: See page 69 of (Fulton 1993). ∎

Formally, the set $\mathcal{A} - \mathbf{a}_\sigma = \{\mathbf{a} - \mathbf{a}_\sigma : \mathbf{a} \in \mathcal{A}\}$ lies in $M \oplus \mathbf{Z}$. But each element has zero last coordinate. We drop it and identify $\mathcal{A} - \mathbf{a}_\sigma$ with its image in M.

Example 13.9. *(Embedding $P^1 \times P^1$ into P^{11} by the line bundle $\mathcal{O}(3,2)$)*
Let Δ be the fan in $\mathbf{Z}^2$ consisting of the four orthants. Then X_Δ equals the product of two projective lines, $P^1 \times P^1$. The four rays of Δ are denoted $D_{(1,0)}, D_{(0,1)}, D_{(-1,0)}, D_{(0,-1)}$. They correspond to the codimension 1 orbits on X_Δ. The following divisor is very ample:

$$D \quad = \quad 2D_{(1,0)} + D_{(0,1)} + D_{(-1,0)} + D_{(0,-1)}.$$

The polytope defined by the divisor D is the rectangle

$$P_D \quad = \quad \{(u_1,u_2) \in \mathbf{Q}^2 \, : \, u_1 \leq 2\,,\ u_2 \leq 1\,,\ -u_1 \leq 1\,,\ -u_2 \leq 1\}.$$

In this example the configuration (13.3) equals

$$\mathcal{A} \quad = \quad \{\,(-1,-1,1),\ (-1,0,1),\ (-1,1,1),\ (0,-1,1),\ (0,0,1),\ (0,1,1),\\ (1,-1,1),\ (1,0,1),\ (1,1,1),\ (2,-1,1),\ (2,0,1),\ (2,1,1)\,\}.$$

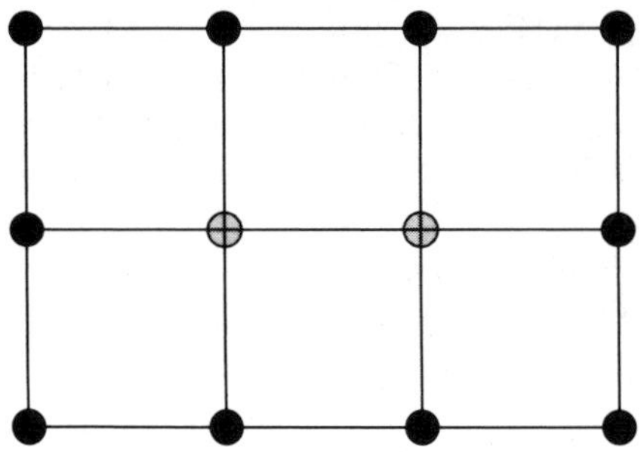

Figure 13-2. Embedding of $P^1 \times P^1$ via $\mathcal{O}(3,2)$.

That the line bundle $\mathcal{O}(D) \simeq \mathcal{O}(3,2)$ is very ample can be seen by verifying the condition in Lemma 13.8. The resulting embedding of $P^1 \times P^1$ into P^{11} has the defining ideal

$$I_{\mathcal{A}} = \langle\, x_2^2 - x_1x_3,\ x_2x_3 - x_1x_4,\ x_2x_5 - x_1x_6,\ \ldots\ ,\ x_{10}x_{11} - x_9x_{12},\ x_{11}^2 - x_{10}x_{12}\,\rangle.$$

Algebraic geometers will find it noteworthy that the sets $\mathcal{A}$ in (13.3) always give embeddings by *complete linear series.* Naturally, it is very well possible to embed a normal projective toric variety X_Δ by an incomplete linear series, corresponding to a certain subset $\mathcal{A}'$ of $\mathcal{A}$. For instance, in Example 13.9 we can take $\mathcal{A}' := \mathcal{A} \setminus \{(0,0,1),(1,0,1)\}$ to get such an embedding of $P^1 \times P^1$ into P^9. The corresponding toric ideal $I_{\mathcal{A}'} = I_{\mathcal{A}} \cap k[x_1,\ldots,x_5,x_8,x_9,\ldots x_{12}]$ is not generated by quadrics (in contrast to Conjecture 13.19 below). Among its 22 minimal generators there are two cubics. Our claim that $Y_{\mathcal{A}'}$ is still isomorphic to $P^1 \times P^1$ can be verified in the previous diagram: removal of the two interior points does not change the semigroups corresponding to the four affine coordinate charts. ■

For the remainder of Chapter 13 we assume that $\mathcal{A} = \{\mathbf{a}_1,\ldots,\mathbf{a}_n\}$ spans $\mathbf{Z}^d$ and is a *graded set* in the sense of Lemma 4.14, meaning that $I_{\mathcal{A}}$ is homogeneous in the usual grading $deg(x_i) = 1$. The toric ideal $I_{\mathcal{A}}$ defines an affine toric variety $X_{\mathcal{A}}$ of dimension d in k^n and a projective toric variety $Y_{\mathcal{A}}$ of dimension $d-1$ in P^{n-1}. The variety $Y_{\mathcal{A}}$ is *projectively normal* if the equivalent conditions of Proposition 13.5 hold. A weaker requirement for $Y_{\mathcal{A}}$ is to be *normal.* Readers of (Hartshorne 1977) will find a normal projective toric variety which is not projectively normal on page 23 in Exercise I.3.18 (b). To address the issue of normality we need to examine the natural affine cover of $Y_{\mathcal{A}}$.

Lemma 13.10. *The projective toric variety $Y_{\mathcal{A}}$ has an open cover consisting of the affine toric varieties $X_{\mathcal{A}-\mathbf{a}_i}$, where $\mathbf{a}_i$ runs over the vertices of the $(d-1)$-polytope $Q = conv(\mathcal{A})$.*

Proof: The image of the homogeneous ideal $I_{\mathcal{A}}$ under the substitution $x_i \mapsto 1$ is the toric ideal $I_{\mathcal{A}-\mathbf{a}_i}$ associated with $\mathcal{A} - \mathbf{a}_i = \{\,\mathbf{a} - \mathbf{a}_i \,:\, \mathbf{a} \in \mathcal{A}\,\}$. Therefore the i-th affine coordinate chart $\{\,x_i \neq 0\,\}$ in P^{n-1} intersects the projective variety $Y_{\mathcal{A}}$ precisely in the affine variety $X_{\mathcal{A}-\mathbf{a}_i}$, and we get an affine covering

$$Y_{\mathcal{A}} \quad = \quad X_{\mathcal{A}-\mathbf{a}_1} \cup X_{\mathcal{A}-\mathbf{a}_2} \cup \cdots \cup X_{\mathcal{A}-\mathbf{a}_n}. \tag{13.4}$$

Let $\mathbf{a}_1, \ldots, \mathbf{a}_s$ be the vertices of $Q = conv(\mathcal{A})$. For each non-vertex $\mathbf{a}_i$, $i = s+1, \ldots, n$, there exists a homogeneous binomial of the form $x_i^{d_i} - \prod_{j=1}^{s} x_j^{d_j}$ in the ideal $I_\mathcal{A}$. Hence all points $\mathbf{x} = (x_1 : \cdots : x_n)$ on $Y_\mathcal{A}$ must have a non-zero entry among its first s coordinates. Equivalently, the affine charts $X_{\mathcal{A}-\mathbf{a}_i}$ for $i > s$ are redundant in the union (13.4). ∎

We recall the definitions of the *Hilbert polynomial* and the *Ehrhart polynomial*:

$$\begin{array}{llll} H_\mathcal{A}(r) & := & \#\{\mathbf{a}_{i_1} + \cdots + \mathbf{a}_{i_r} \, : \, \mathbf{a}_{i_1}, \ldots, \mathbf{a}_{i_r} \in \mathcal{A}\} & \text{for} \quad r \gg 0 \\ E_\mathcal{A}(r) & := & \#(r \cdot conv(\mathcal{A}) \cap \mathbf{Z}\mathcal{A}) & \text{for} \quad r \geq 0. \end{array}$$

It follows directly from the definition that $H_\mathcal{A}(r) \leq E_\mathcal{A}(r)$. In the proof of Theorem 4.16 it was shown that both polynomials have the same initial term

$$\frac{Vol(Q)}{dim(\mathcal{A})!} \cdot r^{dim(\mathcal{A})}.$$

Theorem 13.11. *The projective toric variety $Y_\mathcal{A}$ is normal if and only if its Hilbert polynomial $H_\mathcal{A}$ and its Ehrhart polynomial $E_\mathcal{A}$ are equal.*

Proof: Let $\mathbf{a}_1, \ldots, \mathbf{a}_s$ denote the vertices of $Q = conv(\mathcal{A})$. The projective variety $Y_\mathcal{A}$ is normal if and only if all of its charts $X_{\mathcal{A}-\mathbf{a}_i}$ are normal affine toric varieties. By Proposition 13.5 and Lemma 13.10, the latter condition is equivalent to

$$\mathbf{N}(\mathcal{A} - \mathbf{a}_i) \quad = \quad pos(\mathcal{A} - \mathbf{a}_i) \, \cap \, \mathbf{Z}(\mathcal{A} - \mathbf{a}_i) \qquad \text{for} \ \ i = 1, 2, \ldots, s. \qquad (13.5)$$

To prove the if-direction suppose that (13.5) does not hold. Choose a vector $\mathbf{b}$ in the right hand side of (13.5) which does not lie in the left hand side. For all sufficiently large integers $r \gg 0$, the vector $\mathbf{b} + r\mathbf{a_i}$ lies in $pos(\mathcal{A}) \cap \mathbf{Z}\mathcal{A}$, but it does not lie in $\mathbf{N}\mathcal{A}$. Therefore $H_\mathcal{A}(r) < E_\mathcal{A}(r)$ for $r \gg 0$.

We next prove the only-if-direction. Suppose that (13.5) holds. The *degree* of a vector $\mathbf{b} \in \mathbf{Z}\mathcal{A}$ is the inner product $\omega \cdot \mathbf{b}$, where $\omega \in \mathbf{Q}^d$ is the grading functional in Lemma 4.14. The semigroup algebra of $pos(\mathcal{A}) \cap \mathbf{Z}\mathcal{A}$ is a finitely generated graded module over $k[\mathcal{A}]$. This means that there exists a unique minimal finite set $\mathcal{H} \subset pos(\mathcal{A}) \cap \mathbf{Z}\mathcal{A}$ such that

$$pos(\mathcal{A}) \cap \mathbf{Z}\mathcal{A} \quad = \quad \mathcal{H} + \mathbf{N}\mathcal{A}. \qquad (13.6)$$

We claim that each vector $\mathbf{h} \in \mathcal{H}$ has degree at most $d-1$. Suppose on the contrary that $degree(\mathbf{h}) \geq d$. Choose indices $i_1, \ldots, i_d$ such that $\mathbf{h} \in pos(\{\mathbf{a}_{i_1}, \ldots, \mathbf{a}_{i_d}\})$ and write $\mathbf{h} = \lambda_1 \mathbf{a}_{i_1} + \cdots + \lambda_d \mathbf{a}_{i_d}$, where λ_i are non-negative rationals. Then we have $degree(\mathbf{h}) = \lambda_1 + \cdots + \lambda_d \geq d$, and therefore $\lambda_j \geq 1$ for some j. The vector $\mathbf{h} - \mathbf{a}_{i_j}$ lies in the semigroup $pos(\mathcal{A}) \cap \mathbf{Z}\mathcal{A}$. This is a contradiction to our assumption that $\mathbf{h}$ is a minimal generator of that semigroup over $\mathbf{N}\mathcal{A}$.

Let $\mathbf{h} \in \mathcal{H}$ and $i \in \{1, \ldots, s\}$ and consider the degree 0 vector $\mathbf{h} - (\mathbf{h} \cdot \omega) \cdot \mathbf{a}_i$. It lies in the right hand side of (13.5), and hence it lies in the semigroup $\mathbf{N}(\mathcal{A} - \mathbf{a}_i)$. Adding a sufficiently large positive multiple of $\mathbf{a}_i$ to this vector yields an element of $\mathbf{N}\mathcal{A}$. We conclude that there exists an integer $R > 0$ such that

$$\mathcal{H} + R \cdot \mathbf{a}_i \quad \subset \quad \mathbf{N}\mathcal{A} \qquad \text{for} \ \ i = 1, \ldots, s. \qquad (13.7)$$

Let $\mathbf{b}$ be an arbitrary vector in $pos(\mathcal{A}) \cap \mathbf{Z}\mathcal{A}$ whose degree $\mathbf{b}\cdot\omega$ is at least Rd. There exists a subset $\{i_1, \ldots, i_d\} \subset \{1, 2, \ldots, s\}$ and rational numbers $\lambda_1, \ldots, \lambda_d \geq 0$ such that $\mathbf{b} = \lambda_1 \mathbf{a}_{i_1} + \cdots + \lambda_d \mathbf{a}_{i_d}$. Since $\lambda_1 + \cdots + \lambda_d \geq Rd$, at least one index satisfies $\lambda_{i_j} \geq R$. This implies $\mathbf{b} - R\mathbf{a}_{i_j} \in pos(\mathcal{A})$. Using (13.6) and (13.7) we conclude

$$\mathbf{b} \;\in\; R\mathbf{a}_{i_j} + (pos(\mathcal{A}) \cap \mathbf{Z}\mathcal{A}) \;=\; R\mathbf{a}_{i_j} + \mathcal{H} + \mathbf{N}\mathcal{A} \;\subset\; \mathbf{N}\mathcal{A}.$$

This implies that $E_{\mathcal{A}}(r) = H_{\mathcal{A}}(r)$ for all integers $r \geq Rd$. ∎

There is a natural generalization of Theorem 13.11 which explains the difference between the Ehrhart polynomial and the Hilbert polynomial in geometric terms. A projective variety Y is said to be *normal in dimension* $\geq e$ if, for every irreducible subvariety Y' of dimension $\geq e$, the local ring $\mathcal{O}_{Y,Y'}$ of Y along Y' is integrally closed. If F is a face of the polytope $Q = conv(\mathcal{A})$, then we write $\mathbf{Z}(\mathcal{A}/F)$ for the quotient lattice $\mathbf{Z}\mathcal{A}/\mathbf{Z}(F \cap \mathcal{A})$, and we write $\mathcal{A}/F$ for the image of $\mathcal{A}$ in this lattice. We omit the proof of Theorem 13.12.

Theorem 13.12. *For a graded set $\mathcal{A} = \{\mathbf{a}_1, \ldots, \mathbf{a}_n\} \subset \mathbf{Z}^d$ the following are equivalent:*
(i) The projective toric variety $Y_{\mathcal{A}}$ is normal in dimension $\geq e$;
(ii) $pos(\mathcal{A}/F) \cap \mathbf{Z}(\mathcal{A}/F) = \mathbf{N}(\mathcal{A}/F)$ for every e-dimensional face F of Q;
(iii) The polynomial $E_{\mathcal{A}}(r) - H_{\mathcal{A}}(r)$ has degree less than e.

It is easy to go back from the $\mathcal{A}$-point of view to the setting of algebraic geometry as presented in (Oda 1988) and (Fulton 1993). The normalization of a projective toric variety $Y_{\mathcal{A}}$ is the normal projective toric variety X_Δ defined by the normal fan $\Delta = \mathcal{N}(Q)$ of the polytope $Q = conv(\mathcal{A})$. Here Δ is taken with respect to the lattice $N = (\mathbf{Z}\mathcal{A})^\vee$. If the condition in Theorem 13.11 holds, then $Y_{\mathcal{A}}$ and X_Δ are isomorphic.

If $\mathcal{A}'$ is a subset of $\mathcal{A}$ then there is a natural projection (by deleting coordinates) from $Y_{\mathcal{A}}$ onto $Y_{\mathcal{A}'}$ if and only if $conv(\mathcal{A}) = conv(\mathcal{A}')$. This condition is necessary and sufficient (by Lemma 13.10) for the projection to be everywhere defined. (Birational maps are not considered here.) A typical situation is that $Y_{\mathcal{A}}$ is normal and its image $Y_{\mathcal{A}'}$ is no longer normal. For instance, if $d = 2$, $\mathcal{A} = \{(3,0), (2,1), (1,2), (0,3)\}$ and $\mathcal{A}' = \{(3,0), (1,2), (0,3)\}$, then $Y_{\mathcal{A}} \to Y_{\mathcal{A}'}$ is the map from P^1 onto a cuspidal cubic curve (Hartshorne 1977, Exercise I.3.14 (b), page 22). It is instructive to notice that not every projective toric variety can be gotten by such a coordinate projection.

Example 13.13. *(Saturation may destroy the standard grading)*
We shall present a graded set $\mathcal{A}$ which cannot be extended to a graded set $\mathcal{A}'$ with the same convex hull such that $Y_{\mathcal{A}'}$ is normal. In other words, $Y_{\mathcal{A}}$ is not a proper coordinate projection of any normal projective toric variety $Y_{\mathcal{A}'}$. Let $d = 4, n = 5$ and define

$$\mathcal{A} \;=\; \big\{(0,0,0,1), (0,0,1,1), (1,0,1,1), (0,1,1,1), (1,1,4,1)\big\}.$$

The projective variety $Y_{\mathcal{A}}$ is a hypersurface of degree four in P^4. Its ideal is

$$I_{\mathcal{A}} \;=\; \langle\, x_1^3 x_5 \;-\; x_2^2 x_3 x_4 \,\rangle$$

The polytope $Q = conv(\mathcal{A})$ is a bipyramid, which contains no lattice points from $\mathbf{Z}\mathcal{A} = \mathbf{Z}^4$ other than its vertices. Therefore $Y_\mathcal{A}$ is not a proper coordinate projection of any other projective toric variety. We must show that $Y_\mathcal{A}$ is not normal. Consider the affine chart defined by the vertex $\mathbf{a}_5 = (1,1,4,1)$ of Q. Then $\mathbf{N}(\mathcal{A} - \mathbf{a}_5)$ is isomorphic to the semigroup spanned by $\mathcal{B} = \{(1,1,4), (1,1,3), (0,1,3), (1,0,3)\}$. The vector $(1,1,5)$ lies in $\mathbf{Z}\mathcal{B} \cap pos(\mathcal{B})$ but not in $\mathbf{N}\mathcal{B}$. Therefore $X_\mathcal{B} = X_{\mathcal{A}-\mathbf{a}_5}$ is not normal, and hence $Y_{\mathcal{A}-\mathbf{a}_5}$ is not normal. Indeed, the Hilbert polynomial and the Ehrhart polynomial are different:

$$H_\mathcal{A}(r) \;=\; \frac{2}{3}r^3 + r^2 + \frac{7}{3}r + 1, \qquad \text{while} \quad E_\mathcal{A}(r) \;=\; \frac{2}{3}r^3 + \frac{3}{2}r^2 + \frac{11}{6}r + 1.$$

We see that $Y_\mathcal{A}$ is not normal in codimension 1. The condition (ii) of Theorem 13.12 is violated for the facet F spanned by the last three vectors in $\mathcal{A}$. The isomorphism $\mathbf{Z}(\mathcal{A}/F) \simeq \mathbf{Z}^1$ is defined by the linear form $\mathbf{n} = (-3,-3,1,2)$. The one-dimensional set $\mathcal{A}/F = \{\mathbf{n}\cdot\mathbf{a}_1, \cdots, \mathbf{n}\cdot\mathbf{a}_5\} = \{2,3,0,0,0\}$ does not span a normal semigroup. ∎

Our next result concerns the ideal of a projectively normal toric variety $Y_\mathcal{A}$.

Theorem 13.14. *Let $\mathcal{A} \subset \mathbf{Z}^d$ be a graded set such that $\mathbf{N}\mathcal{A}$ is a normal semigroup. Then the toric ideal $I_\mathcal{A}$ is generated by homogeneous binomials of degree at most d.*

Proof: Choose a generic $n \times n$-matrix U over k and perform the linear change of variables $\mathbf{x} \mapsto U\cdot\mathbf{z}$, We identify $I_\mathcal{A}$ with its image under U. This is a homogeneous ideal in $k[\mathbf{z}] = k[z_1, \ldots, z_n]$. (It is is not a binomial ideal !) Let $\prec$ denote the reverse lexicographic term order with $z_1 \prec z_2 \prec \cdots \prec z_n$. Suppose $dim(\mathcal{A}) = d$. Then $I_\mathcal{A} \cap k[z_1, \ldots, z_d] = \{0\}$, which means that $\{z_1, \ldots, z_d\}$ is a linear system of parameters for $k[\mathcal{A}]$.

We shall apply Hochster's Theorem that $k[\mathcal{A}] = k[\mathbf{z}]/I_\mathcal{A}$ is *Cohen-Macaulay*; see (Oda 1988, Corollary 3.9). Since we are in generic coordinates, the results of Bayer and Stillman (1987a) guarantee that $k[\mathbf{z}]/in_\prec(I_\mathcal{A})$ is Cohen Macaulay as well. Since $in_\prec(I_\mathcal{A})$ is Borel-fixed, this means that $in_\prec(I_\mathcal{A})$ is generated by monomials only in the last $n-d$ variables $z_{d+1}, \ldots, z_n$. There are only finitely many standard monomials in these variables. Let h_j denote the number of monomials $z_{d+1}^{i_{d+1}} \cdots z_n^{i_n}$ not in $in_\prec(I_\mathcal{A})$ of degree $j = i_{d+1} + \cdots + i_n$. The common Hilbert series of $k[\mathcal{A}] = k[\mathbf{z}]/I_\mathcal{A}$ and $k[\mathbf{z}]/in_\prec(I_\mathcal{A})$ equals

$$\sum_{r=0}^{\infty} E_\mathcal{A}(r)\cdot z^r \quad = \quad \frac{h_0 + h_1 z + h_2 z^2 + \cdots + h_s z^s}{(1-z)^d}. \tag{13.8}$$

Here $E_\mathcal{A}(r)$ is the Ehrhart polynomial of the $(d-1)$-polytope $conv(\mathcal{A})$. This is a polynomial of degree $d-1$. Standard arguments about rational generating functions (Stanley 1986) imply that the numerator polynomial in (13.8) has the same degree $s = d-1$. Hence there are no standard monomials $z_{d+1}^{i_{d+1}} \cdots z_n^{i_n}$ of degree d and higher. This means that $in_\prec(I_\mathcal{A})$ is generated by monomials of degree $\leq d$. We have shown that the reverse lexicographic Gröbner basis of $I_\mathcal{A}$ in generic coordinates consists of forms of degree $\leq d$. ∎

We remark that the bound in Theorem 13.14 is tight. For instance, the graded set

$$\mathcal{A} = \{ d\mathbf{e}_1, d\mathbf{e}_2, \ldots, d\mathbf{e}_d, \mathbf{e}_1 + \mathbf{e}_2 + \cdots + \mathbf{e}_d \}$$

spans a normal semigroup (check this), while its toric ideal is generated in degree d:

$$I_{\mathcal{A}} \quad = \quad \langle x_1 x_2 \cdots x_d - x_{d+1}^d \rangle.$$

Also note that our proof of Theorem 13.14 actually proves a stronger statement, namely, that the toric ideal $I_{\mathcal{A}}$ has *regularity* at most d. It is an open problem whether there exists a Gröbner basis of degree $\leq d$ in the original coordinates $\mathbf{x}$ rather than in the generic coordinates $\mathbf{z}$. Here is the precise question: *If $\mathcal{A}$ satisfies the hypotheses of Theorem 13.14, then does the toric ideal $I_{\mathcal{A}} \subset k[\mathbf{x}]$ possess a Gröbner basis of degree at most d ?* The answer is "yes" if $\mathcal{A}$ admits a unimodular regular triangulation.

Proposition 13.15. *Let $\mathcal{A}$ be a graded subset of $\mathbf{Z}^d$. Suppose that, for some term order $\prec$ on $k[\mathbf{x}]$, the initial monomial ideal $in_\prec(I_{\mathcal{A}})$ is square free. Then*
(i) $Y_{\mathcal{A}}$ is projectively normal, and
(ii) the reduced Gröbner basis of $I_{\mathcal{A}}$ with respect to $\prec$ has degree $\leq d$.

Proof: Let $\Delta_\prec$ be the regular triangulation of $\mathcal{A}$ having Stanley-Reisner ideal $in_\prec(I_{\mathcal{A}})$. By Corollary 8.9 every simplicial cone σ of $\Delta_\prec$ is spanned by a lattice basis $\{\mathbf{a}_{\sigma_1}, \ldots, \mathbf{a}_{\sigma_d}\}$. Every vector $\mathbf{b} \in pos(\mathcal{A})$ lies in one of these cones σ, so that $\mathbf{b} = \lambda_1 \mathbf{a}_{\sigma_1} + \cdots + \lambda_d \mathbf{a}_{\sigma_d}$ where $\lambda_1, \ldots, \lambda_d$ are unique non-negative reals. If in addition $\mathbf{b}$ lies in $\mathbf{Z}\mathcal{A}$, then the coefficients λ_i must be integers, and therefore $\mathbf{b} \in \mathbf{N}\mathcal{A}$. This shows that the semigroup $\mathbf{N}\mathcal{A}$ is normal, which means that $Y_{\mathcal{A}}$ is projectively normal.

For the assertion (ii) we must show that $in_\prec(I_{\mathcal{A}})$ is generated in degree $\leq d$. Every minimal generator of $in_\prec(I_{\mathcal{A}})$ is a square-free monomial $x_{i_1} x_{i_2} \cdots x_{i_s}$ such that $\{i_1, \ldots, i_s\}$ is not a face of Δ but each of its proper subsets is a face of Δ. Since Δ triangulates the polytope $conv(\mathcal{A})$, it is a pure $(d-1)$-dimensional contractible simplicial complex. The boundary of the $(s-1)$-simplex $\{i_1, \ldots, i_s\}$ lies in Δ, and it must therefore be the boundary of an $(s-1)$-dimensional subcomplex of Δ. This implies $s \leq d$. ∎

Proposition 13.15 is a tool for showing that a toric variety is projectively normal.

Example 13.16. *(Toric varieties in the Grassmann variety of lines in P^{d-1})*
Let ξ be a generic point on the Grassmann variety $Grass_{2,d}$ of 2-dimensional linear subspaces of k^d. Let $Y := \overline{(k^*)^d \cdot \xi}$ denote its orbit closure under the natural action of the torus $(k^*)^d$. We shall prove that Y is projectively normal. To this end we consider the Plücker embedding of $Grass_{2,d}$ into the projectivization of $\wedge_2 k^d$. The image of Y under the Plücker embedding is isomorphic (in the $\mathcal{A}$-graded sense of Chapter 10) to the toric variety $Y_{\mathcal{A}}$, where $\mathcal{A} = \{e_i + e_j : 1 \leq i < j \leq d\}$ is the set of weights of $\wedge_2 k^d$ as a $GL(d)$-module. It was shown in Theorem 9.1 that the toric ideal $I_{\mathcal{A}}$ possesses a square-free initial ideal. Using Proposition 13.15, we conclude that $Y = Y_{\mathcal{A}}$ is projectively normal. In Chapter 14 we shall see that our square-free quadratic Gröbner basis generalizes to the generic toric variety in the Grassmann variety $Grass_{r,d}$ of r-dimensional subspaces. ∎

Unfortunately this technique for showing projective normality does not always apply. Our next example shows that the converse of Proposition 13.15 (i) is false in general.

Example 13.17.
(A projectively normal toric variety without square-free initial ideal)
Let $d = 5, n = 9$ and let $\mathcal{A}$ be the graded set consisting of the columns of the matrix

$$\mathcal{A} \quad = \quad \begin{pmatrix} 0 & 1 & 0 & 0 & 1 & 1 & 1 & 1 & 1 \\ 0 & 0 & 1 & 0 & 1 & 1 & 2 & 2 & 2 \\ 0 & 0 & 0 & 1 & 1 & 2 & 2 & 3 & 3 \\ 0 & 0 & 0 & 0 & 1 & 2 & 3 & 4 & 5 \\ 1 & 1 & 1 & 1 & 1 & 1 & 1 & 1 & 1 \end{pmatrix}.$$

This defines a 4-dimensional projectively normal toric variety of degree 18 in P^8. The normality of the semigroup $\mathbf{N}\mathcal{A}$ was verified by computer (see Exercise (6) below). The toric ideal $I_\mathcal{A}$ has 14 minimal generators: one quadric, twelve cubics and one quartic. We ran the MAPLE program PUNTOS by Jesus De Loera (1995b) to enumerate all regular triangulations of the set $\mathcal{A}$. There are 180 of them and none is unimodular. Using Corollary 8.9 we conclude that no initial monomial ideal of $I_\mathcal{A}$ is square-free. We conjecture the stronger statement that no initial monomial ideal of $I_\mathcal{A}$ is Cohen-Macaulay. (This conjecture is meant with respect to the original coordinates $x_1, \ldots, x_9$, of course.) ∎

The phenomenon of Example 13.17 can be circumvented by enlarging the generating set of the given semigroup. Indeed, the familiar desingularization construction for toric varieties (Fulton 1993, Section 2.6) implies the following result: A semigroup M is normal if and only if there exists a set $\mathcal{A}$ of generators of M which admits a unimodular triangulation. It is noteworthy that the converse of Proposition 13.15 (i) is true for toric surfaces.

Proposition 13.18. *Let $Y_\mathcal{A}$ be a projectively normal toric surface in P^{n-1} defined by the graded subset $\mathcal{A} = \{\mathbf{a}_1, \ldots, \mathbf{a}_n\}$ of $\mathbf{Z}^3$. Order the variables $x_1, x_2, \ldots, x_n$ so that $\mathbf{a}_i \notin conv(\{\mathbf{a}_1, \ldots, \mathbf{a}_{i-1}\})$ for $i = 2, \ldots, n$ and let $\prec$ be the resulting purely lexicographic term order on $k[\mathbf{x}]$. Then $in_\prec(I_\mathcal{A})$ is square-free and generated in degree $d \leq 3$.*

Proof: The polygon $Q = conv(\mathcal{A})$ has the property that each lattice point of Q lies in $\mathcal{A}$. Consider the lexicographic triangulation $\Delta_\prec$ of Q. According to Proposition 8.6, it can be constructed as follows: Proceed recursively and first triangulate $Q_{i-1} := conv(\{\mathbf{a}_1, \ldots, \mathbf{a}_{i-1}\})$. Then join the new point $\mathbf{a}_i$ to every boundary segment of the triangulated polygon Q_{i-1} which is visible from $\mathbf{a}_i$. Each triangle σ in the resulting triangulation $\Delta_\prec$ contains no lattice points except for its three vertices. This implies that the normalized area of σ is one. (The analogous property does not hold in higher dimensions; cf. Example 13.13). Hence $\Delta_\prec$ is a unimodular triangulation and $in_\prec(I_\mathcal{A})$ is a square-free monomial ideal. It is generated in degree ≤ 3 by part (ii) of Proposition 13.15. ∎

The following diagram illustrates the square-free lexicographic Gröbner basis constructed in Proposition 13.18 for the toric ideal $I_\mathcal{A}$ in Example 13.9.

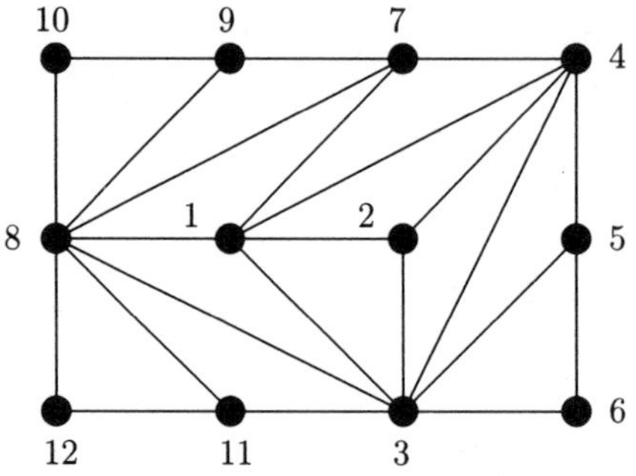

Figure 13-3. Square-free lexicographic Gröbner basis of degree ≤ 3.

One might ask whether the degree bound in Theorem 13.14 can be further improved by putting geometric restrictions, such as smoothness, on the variety $Y_{\mathcal{A}}$.

Conjecture 13.19. *Let $Y_{\mathcal{A}}$ be a projectively normal nonsingular toric variety. Then the toric ideal $I_{\mathcal{A}}$ is generated by quadratic binomials.*

In 1994 Rikard Bøgvad announced this conjecture as a theorem, but unfortunately a gap was found in his proof. Bøgvad's approach is based on the technique of Frobenius splitting in characteristic p and some intersection-theoretic arguments. It is beyond the scope of this monograph. Note, however, that both the hypothesis and the conclusion of Conjecture 13.19 are combinatorial statements. (Recall that $Y_{\mathcal{A}}$ is smooth if and only if the semigroup $\mathbf{N}(\mathcal{A} - \mathbf{a})$ is isomorphic to $\mathbf{N}^{d-1}$ for every vertex $\mathbf{a}$ of $conv(\mathcal{A})$.) So, we may hope for a purely combinatorial proof of Conjecture 13.19. For a special class of toric manifolds such a proof was given by Ewald & Schmeinck (1993). It is also unknown whether the toric ideal $I_{\mathcal{A}}$ of a projectively normal *nonsingular* toric variety $Y_{\mathcal{A}}$ always has a quadratic Gröbner basis. We remark that this does not hold in *generic coordinates.* They were denoted $\mathbf{z} = (z_1, \ldots, z_n)$ in the proof of Theorem 13.14. As an example we take $Y_{\mathcal{A}}$ to be the cubic Veronese embedding of P^2 into P^8. The regularity of its ideal $I_{\mathcal{A}}$ is three. Every initial ideal of $I_{\mathcal{A}}$ in $k[\mathbf{z}] = k[z_1, \ldots, z_9]$ is Borel-fixed and has regularity at least three and must therefore possess a minimal monomial generator of degree ≥ 3.

Exercises:

(1) In Lemma 13.4 we saw that $X_{\mathcal{A}} \cap (k^*)^n$ is a group.
   (a) What is the identity element of this group ?
   (b) Determine the tangent space of $X_{\mathcal{A}}$ at this identity point.

(2) Give a geometric description of the subdivision $\Delta_{-\mathbf{e}}(\mathcal{A})$ considered in Remark 13.7. Derive a criterion for the tangent cone of a toric singularity to be irreducible.

(3) Consider two graded sets $\mathcal{A}, \mathcal{A}' \subset \mathbf{Z}^d$ and form their sum $\mathcal{A} + \mathcal{A}'$. What is the geometric relationship among the three projective varieties $Y_{\mathcal{A}}$, $Y_{\mathcal{A}'}$ and $Y_{\mathcal{A}+\mathcal{A}'}$? Describe the fans of their normalizations. How is the toric ideal $I_{\mathcal{A}+\mathcal{A}'}$ related to $I_{\mathcal{A}}$ and $I_{\mathcal{A}'}$? What can you say about minimal generators and Gröbner bases ?

(4) Show that every projective toric surface $Y_{\mathcal{A}}$ is a proper coordinate projection of a normal projective toric surface. In other words, the phenomenon of Example 13.13 cannot happen for $d \leq 3$.

(5) Let $\mathcal{A} \subset \mathbf{Z}^d$ be a graded set such that $conv(\mathcal{A}) \cap \mathbf{Z}\mathcal{A} = \mathcal{A}$. Does there exist a bound in terms of d for the degrees of the minimal generators of $I_{\mathcal{A}}$?

(6) Let $\mathcal{A}$ be the configuration in Example 13.17.
  (a) Compute the Hilbert polynomial $H_{\mathcal{A}}(r)$.
  (b) Show that $H_{\mathcal{A}}(r)$ coincides with the *Hilbert function* $r \mapsto dim_k(k[\mathbf{x}]/I_{\mathcal{A}})_r$ for <u>all</u> non-negative integers r.
  (c) Compute the Ehrhart polynomial $E_{\mathcal{A}}(r)$.
  (d) Conclude that the projective toric variety $Y_{\mathcal{A}}$ is projectively normal.

(7) Let $I_{\mathcal{A}}$ be the ideal of the cubic Veronese embedding of P^2 into P^8, considered in generic coordinates $\mathbf{z} = (z_1, \ldots, z_9)$ as in the proof of Theorem 13.14.
  (a) Show that $k[\mathcal{A}]$ is a free module of rank 9 over $k[z_1, z_2, z_3]$.
  (b) Determine an explicit module basis.
  (c) Express the element $(z_1 + \cdots + z_9)^4$ in your basis.

(8) Let $Y_{\mathcal{A}}$ be a projectively normal toric surface. Show that the toric ideal $I_{\mathcal{A}}$ is generated by quadrics if and only if the polygon $Q = conv(\mathcal{A})$ has more than three lattice points on its boundary. This result is due to Koelman (1993b).

(9) Give an algorithm for computing the normalization of an affine toric variety $X_{\mathcal{A}}$.

(10) Write down explicitly the reduced Gröbner basis defined by Figure 13-3.

Notes:
Some exciting current research on toric varieties is motivated by the mirror symmetry conjecture of mathematical physics. These developments involve surprising connections to the material presented in this monograph. Two examples are the Gröbner basis for the quantum cohomology ring of a toric manifold given by Batyrev (1993) and the role of the secondary fan $\Sigma(\mathcal{A})$ in the work of Aspinwall, Greene and Morrison (1993).

Theorem 13.14 is a folklore result of commutative algebra. It was first explained to me by Victor Batyrev in 1992. For the special case of toric surfaces there is a nice combinatorial proof of Theorem 13.14 in (Koelman 1993a). The construction in Example 13.16 is based on an example of Bouvier and Gonzalez-Sprinberg (1992), which was shown to me in a simplified form by Günter Ziegler.

The normality result in Example 13.16 is by no means best possible. Stronger results were proved using different methods by White (1977) and Dabrowski (1996). White shows that the basis monomial ring of a matroid is normal, which implies that every torus orbit closure in a Grassmann variety (not just the generic one) is projectively normal. Dabrowski shows that generic torus orbit closures in any flag variety G/P are normal.

CHAPTER 14

Some Specific Gröbner Bases

There is an abundance of toric varieties $X_{\mathcal{A}}$ arising naturally in combinatorics, in geometry and in the applications of Chapter 5. Each of the underlying configurations $\mathcal{A}$ calls for a project to examine its toric ideal $I_{\mathcal{A}}$, its Gröbner bases, its syzygies, its triangulations, its state polytope, etc... This was illustrated in Chapter 9 for the second hypersimplex. In this chapter we discuss Gröbner bases for three special families of toric ideals. The first family of configurations to be examined is a common generalization of hypersimplices and Veronese embeddings of projective space. Our second configuration $\mathcal{A}$ lives in matrix space: it is the set of $r \times r$-permutation matrices. Our third configuration is a higher-dimensional generalization of Example 5.1 (contingency tables and transportation problems).

14.A. A square-free quadratic Gröbner basis for varieties of Veronese-type. We fix positive integers r and $s_1, \ldots, s_d$, and we consider the set

$$\begin{aligned} \mathcal{A} \quad = \quad \big\{(i_1, i_2, \ldots, i_d) \in \mathbf{Z}^d \, : i_1 + \cdots + i_d = r, \\ 0 \le i_1 \le s_1, \ldots, \; 0 \le i_d \le s_d \big\} \end{aligned} \qquad (14.1)$$

This family of toric ideals includes many familiar examples. For $s_1 = \cdots = s_d = r$ the toric variety $X_{\mathcal{A}}$ is the *r-th Veronese embedding* of P^{d-1}. For $s_1 = \cdots = s_d = 1$ the polytope $Q = conv(\mathcal{A})$ is the r-th hypersimplex of dimension $d-1$, and the corresponding toric variety is naturally embedded in the Grassmann variety of r-dimensional linear subspaces of k^d (as in Example 13.16). The second hypersimplex $(r = 2, s_1 = \cdots = s_d = 1)$ was discussed in detail in Chapter 9. In what follows we shall present a square-free, quadratic Gröbner basis for $I_{\mathcal{A}}$ which generalizes our "thrackle construction" in Theorem 9.1.

There is a natural bijection between the elements of $\mathcal{A}$ and weakly increasing strings of length r over the alphabet $\{1, 2, \ldots, d\}$ having at most s_j occurrences of the letter j. Under this bijection, the vector $(i_1, i_2, i_3, \ldots, i_d) \in \mathcal{A}$ is mapped to the weakly increasing string

$$u_1 u_2 \cdots u_r \quad = \quad \underbrace{11\cdots1}_{i_1 \text{ times}} \underbrace{22\cdots2}_{i_2 \text{ times}} \underbrace{33\cdots3}_{i_3 \text{ times}} \cdots \underbrace{dd\cdots d}_{i_d \text{ times}}.$$

We write $x_{u_1 u_2 \ldots u_r}$ for the corresponding variable in our polynomial ring $k[\mathbf{x}]$. Let $sort(\,\cdot\,)$ denote the operator which takes any string over the alphabet $\{1, 2, \ldots, d\}$ and sorts it into weakly increasing order. With these convention our toric ideal is described as follows:

Remark 14.1. *The toric ideal defined by the set (14.1) equals*

$$I_{\mathcal{A}} \quad = \quad \langle\, x_{\mathbf{u}} x_{\mathbf{v}} \cdots x_{\mathbf{w}} - x_{\mathbf{u}'} x_{\mathbf{v}'} \cdots x_{\mathbf{w}'} \;:\; sort(\mathbf{u}\mathbf{v}\cdots\mathbf{w}) = sort(\mathbf{u}'\mathbf{v}'\cdots\mathbf{w}')\,\rangle.$$

For instance, the ideal of the Veronese surface in P^5 equals in this notation

$$\langle \underline{x_{11}x_{33}} - x_{13}^2, \underline{x_{11}x_{22}} - x_{12}^2, \underline{x_{11}x_{23}} - x_{12}x_{13},$$
$$\underline{x_{12}x_{33}} - x_{13}x_{23}, \underline{x_{13}x_{22}} - x_{12}x_{23}, \underline{x_{33}x_{22}} - x_{23}^2\rangle.$$

In fact, these six minimal generators constitute the reduced Gröbner basis for the Veronese ideal with respect to any term order which selects the underlined initial terms. This the special case $d = 3, r = s_1 = s_2 = s_3 = 2$ of Theorem 14.2 below.

A monomial $x_{u_1u_2\cdots u_r}x_{v_1v_2\cdots v_r}\cdots x_{w_1w_2\cdots w_r}$ in $k[\mathbf{x}]$ is said to be *sorted* if

$$u_1 \leq v_1 \leq \cdots \leq w_1 \leq u_2 \leq v_2 \leq \cdots \leq w_2 \leq u_3 \leq \cdots \leq u_d \leq v_d \leq \cdots \leq w_d. \tag{14.2}$$

For monomials which are not sorted we define the *inversion number* to be the number of inversions in the string (14.2). By an *inversion* in a string of integers $\ell_1\ell_2\cdots\ell_\nu$ we mean a pair of indices (i, j) such that $i < j$ and $\ell_i > \ell_j$. The following two facts are easily verified:

(i) Every power of a variable is sorted.
(ii) If a monomial is not sorted, then it contains a quadratic factor which is not sorted.

Theorem 14.2. *There exists a term order $\prec$ on $k[\mathbf{x}]$ such that the sorted monomials are precisely the $\prec$-standard monomials modulo $I_\mathcal{A}$. The initial ideal $in_\prec(I_\mathcal{A})$ is generated by square-free quadratic monomials. The corresponding reduced Gröbner basis of $I_\mathcal{A}$ equals*

$$\{ \underline{x_{u_1\cdots u_r}x_{v_1\cdots v_r}} - x_{w_1w_3\cdots w_{2r-1}}x_{w_2w_4\cdots w_{2r}} :$$
$$w_1w_2w_3\cdots w_{2r} = sort(u_1v_1u_2v_2\cdots u_rv_r)\}$$

Proof: Let $\mathcal{G}$ denote the above set of marked binomials

$$\underline{x_{u_1\cdots u_r}x_{v_1\cdots v_r}} - x_{w_1w_3\cdots w_{2r-1}}x_{w_2w_4\cdots w_{2r}} \tag{14.3}$$

We first show that these relations do indeed lie in $I_\mathcal{A}$. Note that for each $j \in \{1, \ldots, d\}$ the strings $u_1\cdots u_r$ and $v_1\cdots v_r$ each have at most s_j occurrences of the letter j. We must verify that the strings $w_1w_3\cdots w_{2r-1}$ and $w_2w_4\cdots w_{2r}$ have the same property. This holds because the number of j's in $w_1w_3\cdots w_{2r-1}$ and the number of j's in $w_2w_4\cdots w_{2r}$ are either equal or they differ by one, which in turn follows from $w_1 \leq w_2 \leq \cdots \leq w_{2r}$.

Consider the reduction relation on $k[\mathbf{x}]$ defined by the marked binomials (14.3). A monomial m is in normal form with respect to this reduction relation if and only if m is sorted (this was observed in (ii) above). If a non-sorted monomial m_1 is reduced to another monomial m_2 using $\mathcal{G}$, then the inversion number of m_2 is strictly less than the inversion number of m_1. This shows that the reduction relation defined by $\mathcal{G}$ is Noetherian. By Theorem 3.12, this implies that the given marking is coherent: there exists a term order $\prec$ on $k[\mathbf{x}]$ which selects the underlined term as the initial term for each binomial in $\mathcal{G}$.

Consider the initial ideal $in_\prec(I_\mathcal{A})$. Every non-sorted monomial lies in this ideal. Suppose that some sorted monomial m_1 lies in $in_\prec(I_\mathcal{A})$. There exists a non-zero binomial $m_1 - m_2 \in I_\mathcal{A}$ such that m_2 does not lie in $in_\prec(I_\mathcal{A})$. Then m_2

is sorted as well, that is, m_1 and m_2 are sorted monomials which lie in the same residue class modulo $I_{\mathcal{A}}$. It follows from the description of $I_{\mathcal{A}}$ in Remark 14.1 that m_1 and m_2 are equal. This is a contradiction. Hence the monomials in $in_{\prec}(I_{\mathcal{A}})$ are precisely the non-sorted monomials. We conclude that the set $\mathcal{G}$ equals the reduced Gröbner basis of $I_{\mathcal{A}}$ with respect to $\prec$. ∎

The above construction generalizes the thrackles in Chapter 9.

Remark 14.3. *In the special case of the second hypersimplex ($r = 2, s_1 = \cdots = s_d = 1$), the Gröbner basis in Theorem 14.2 coincides with the Gröbner basis in Theorem 9.1.*

Proof: The characterization of standard monomials in (9.3) is "to be sorted". ∎

There is an important difference between the special case of the second hypersimplex and the general case. The thrackle triangulation of the second hypersimplex was seen to be lexicographic in Remark 9.2. This does not hold in general.

Proposition 14.4. *In general, the reduced Gröbner basis given in Theorem 14.2 is neither lexicographic nor reverse lexicographic.*

Proof: We list our Gröbner basis explicitly in the case $d = 4$, $r = 3$, and $s_1 = s_2 = s_3 = s_4 = 2$. Then

$$\begin{aligned}
\mathcal{A} =& \{(2,1,0,0),(1,2,0,0),\ldots,(0,0,1,2),(1,1,1,0),(1,1,0,1),(1,0,1,1),(0,1,1,1)\},\\
k[\mathbf{x}] =& [x_{112}, \quad x_{122}, \quad \ldots, \quad x_{344}, \quad x_{123}, \quad x_{124}, \quad x_{134}, \quad x_{234}].
\end{aligned}$$

Hence $k[\mathbf{x}]$ is a polynomial ring in 16 variables. Our Gröbner basis equals $\mathcal{G} =$

$$\begin{aligned}
\{\ & \underline{x_{112}}x_{133} - x_{113}x_{123},\ x_{112}x_{134} - \underline{x_{113}}x_{124},\ x_{112}x_{144} - \underline{x_{114}}x_{124},\ x_{112}x_{223} - \underline{x_{122}}x_{123},\\
& x_{112}x_{224} - x_{122}x_{124},\ x_{112}x_{233} - x_{123}^2,\ x_{112}x_{234} - x_{123}x_{124},\ x_{112}x_{244} - x_{124}^2,\\
& x_{112}x_{334} - x_{123}x_{134},\ x_{112}x_{344} - x_{124}x_{134},\ \underline{x_{113}}x_{122} - x_{112}x_{123},\ x_{113}x_{144} - x_{114}x_{134},\\
& x_{113}x_{223} - x_{123}^2,\ x_{113}x_{224} - x_{123}x_{124},\ x_{113}x_{233} - x_{123}x_{133},\ x_{113}x_{234} - x_{123}x_{134},\\
& x_{113}x_{244} - x_{124}x_{134},\ x_{113}x_{334} - x_{133}x_{134},\ x_{113}x_{344} - x_{134}^2,\ \underline{x_{114}}x_{122} - \underline{x_{112}}x_{124},\\
& x_{114}x_{123} - x_{113}x_{124},\ x_{114}x_{133} - x_{113}x_{134},\ x_{114}x_{223} - x_{123}x_{124},\ x_{114}x_{224} - x_{124}^2,\\
& x_{114}x_{233} - x_{123}x_{134},\ x_{114}x_{234} - x_{124}x_{134},\ x_{114}x_{244} - x_{124}x_{144},\ x_{114}x_{334} - x_{134}^2,\\
& x_{114}x_{344} - x_{134}\underline{x_{144}},\ \underline{x_{122}}x_{133} - x_{123}^2,\ x_{122}x_{134} - x_{123}x_{124},\ x_{122}x_{144} - x_{124}^2,\\
& x_{122}x_{233} - x_{123}\underline{x_{223}},\ x_{122}x_{234} - x_{123}\underline{x_{224}},\ x_{122}x_{244} - x_{124}x_{224},\ x_{122}x_{334} - \underline{x_{123}}x_{234},\\
& x_{122}x_{344} - x_{124}x_{234},\ \underline{x_{123}}x_{144} - x_{124}x_{134},\ x_{123}x_{244} - \underline{x_{124}}x_{234},\ x_{123}x_{334} - x_{133}x_{234},\\
& x_{123}x_{344} - x_{134}x_{234},\ \underline{x_{124}}x_{133} - x_{123}x_{134},\ x_{124}x_{223} - x_{123}x_{224},\ x_{124}x_{233} - x_{123}x_{234},\\
& x_{124}\underline{x_{334}} - x_{134}x_{234},\ x_{124}\underline{x_{344}} - x_{134}x_{244},\ \underline{x_{133}}x_{144} - x_{134}^2,\ x_{133}x_{223} - x_{123}\underline{x_{233}},\\
& x_{133}x_{224} - x_{123}x_{234},\ x_{133}x_{244} - x_{134}x_{234},\ x_{133}x_{344} - x_{134}x_{334},\ \underline{x_{134}}x_{223} - x_{123}x_{234},\\
& x_{134}x_{224} - x_{124}x_{234},\ x_{134}x_{233} - \underline{x_{133}}x_{234},\ \underline{x_{144}}x_{223} - x_{124}x_{234},\ x_{144}x_{224} - x_{124}x_{244},\\
& x_{144}x_{233} - \underline{x_{134}}x_{234},\ x_{144}\underline{x_{234}} - x_{134}x_{244},\ x_{144}x_{334} - x_{134}x_{344},\ \underline{x_{223}}x_{244} - x_{224}x_{234},\\
& x_{223}x_{334} - x_{233}\underline{x_{234}},\ x_{223}x_{344} - x_{234}^2,\ \underline{x_{224}}x_{233} - x_{223}x_{234},\ x_{224}x_{334} - x_{234}^2,\\
& x_{224}x_{344} - x_{234}\underline{x_{244}},\ \underline{x_{233}}x_{244} - x_{234}^2,\ x_{233}x_{344} - x_{234}\underline{x_{334}},\ \underline{x_{244}}x_{334} - x_{234}\underline{x_{344}}\ \}.
\end{aligned}$$

In each of these 68 binomials the first term is the initial term and the second term is the trailing term. Note that the trailing terms are sorted while the initial terms are not sorted.

If $\mathcal{G}$ were the Gröbner basis for a lexicographic term order, then there would exist a variable x_{ijk} which appears only in the initial terms. If $\mathcal{G}$ were the Gröbner basis for a reverse lexicographic term order, then there would exist a variable x_{ijk} which appears only in the trailing terms. However, each of the 16 variables appears both in some initial term and in some some trailing term. This is indicated by the underlinings. We conclude that the Gröbner basis $\mathcal{G}$ is neither lexicographic nor reverse lexicographic. ∎

The property of Proposition 14.4 holds even for the third hypersimplex

$$\Delta(3,6) \;=\; conv\,\{e_i + e_j + e_k \,:\, 1 \leq i < j < k \leq 6\}.$$

Here $d = 6, r = 3, s_1 = s_2 = \cdots = s_6 = 1$. The corresponding toric ideal $I_{\mathcal{A}}$ is the kernel of the k-algebra homomorphism

$$k\big[\,x_{rst} \,:\, 1 \leq i < j < k \leq 6\,\big] \;\rightarrow\; k[t_1,t_2,t_3,t_4,t_5,t_6], \;\; x_{ijk} \;\mapsto\; t_i t_j t_k.$$

Here the reduced Gröbner basis $\mathcal{G}$ consists of 69 binomials. Each variable x_{ijk} appears in some initial term, so that $\mathcal{G}$ cannot be reverse lexicographic. The six variables $x_{123}, x_{234}, x_{345}, x_{456}, x_{156}, x_{126}$ appear only in initial terms, so they can come first in a possible lexicographic order. Let X denote the set of the other 14 variables. Our argument shows that the elimination ideal $I \cap k[X]$ has the induced Gröbner basis

$$\begin{aligned} \mathcal{G} \cap k[X] \;\;=\;\; & \big\{x_{124}x_{256} - \underline{x_{125}}x_{246}, x_{124}x_{346} - \underline{x_{134}}x_{246}, \\ & x_{124}x_{356} - \underline{x_{135}}x_{246}, x_{125}x_{134} - \underline{x_{124}}x_{135}, x_{125}x_{346} - x_{135}x_{246}, x_{125}x_{356} - x_{135}\underline{x_{256}}, \\ & x_{134}x_{256} - x_{135}\underline{x_{246}}, x_{134}x_{356} - x_{135}\underline{x_{346}}, x_{136}x_{145} - x_{135}\underline{x_{146}}, x_{136}x_{235} - x_{135}\underline{x_{236}}, \\ & x_{136}x_{245} - x_{135}x_{246}, x_{145}x_{235} - x_{135}\underline{x_{245}}, x_{145}x_{236} - x_{135}x_{246}, x_{146}x_{235} - x_{135}x_{246}, \\ & x_{146}x_{236} - \underline{x_{136}}x_{246}, x_{146}x_{245} - \underline{x_{145}}x_{246}, x_{236}x_{245} - \underline{x_{235}}x_{246}, x_{256}x_{346} - x_{246}\underline{x_{356}}\big\}. \end{aligned}$$

Each of the remaining 14 variables appears in some trailing term and hence cannot be the next variable in a possible lexicographic term order. This shows that the Gröbner basis $\mathcal{G}$ for the hypersimplex $\Delta(3,6)$ is neither lexicographic nor reverse lexicographic.

Each of the support sets $\mathcal{A}$ appearing in (14.1) occurs naturally in the representation theory of the general linear group $GL(d)$. We recall that each irreducible $GL(d)$-module V_λ is indexed by a partition $\lambda = (\lambda_1, \lambda_2, \ldots, \lambda_t)$, where $\lambda_1 \geq \lambda_2 \geq \cdots \geq \lambda_t$. The vector λ is called the *highest weight* of the representation V_λ. The torus $(k^*)^d$ acts on V_λ as the group of diagonal matrices in $GL(d)$. Let $\mathcal{A}_\lambda$ denote the set of all weights occurring in V_λ. Then the toric variety $X_{\mathcal{A}_\lambda}$ is isomorphic to the closure of a generic $(k^*)^n$-orbit in V_λ. We call a partition λ *almost rectangular* if $\lambda_1 = \lambda_2 = \cdots = \lambda_{t-1} \geq \lambda_t$.

Proposition 14.5. *Let λ be an almost rectangular partition. Then the set $\mathcal{A}_\lambda$ of weights appearing in the irreducible $GL(d)$-module V_λ is equal to the set $\mathcal{A}$ in (14.1) with $s_1 = \cdots = s_d = \lambda_1$ and $r = |\lambda| = \sum_{i=1}^t \lambda_i$.*

Proof: This follows from the fact that the weight defined by a partition μ appears in the irreducible representation V_λ if and only if $|\lambda| = |\mu|$ and λ *dominates* μ, i.e., $\sum_{i=1}^j \lambda_i \geq \sum_{i=1}^j \mu_i$ for all j. Since λ is almost rectangular, these inequalities are equivalent to $\mu_1 \leq s_1, \mu_2 \leq s_2, \ldots, \mu_d \leq s_d$, which means that μ lies in $\mathcal{A}$. This proves that $\mathcal{A}_\lambda = \mathcal{A}$. ∎

Examples:

(a) The almost rectangular partition $\lambda = (2,0,0)$ defines the symmetric power $V_\lambda = S^2k^3$. The corresponding toric variety $X_{\mathcal{A}_\lambda}$ is the Veronese surface in P^5.

(b) The almost rectangular partition $\lambda = (1,1,1,0,0,0)$ defines the exterior power $V_\lambda = \wedge^3 k^6$. The corresponding polytope $conv(\mathcal{A}_\lambda)$ is the third hypersimplex $\Delta(3,6)$.

(c) The almost rectangular partition $\lambda = (2,1,0,0)$ defines an irreducible $GL(4)$-module of dimension 20. This can be seen by substituting $t_i \mapsto 1$ in the Schur polynomial

$$\begin{aligned} S_\lambda(t_1,t_2,t_3,t_4) = {} & t_1^2t_2 + t_1^2t_3 + t_1^2t_4 + t_1t_2^2 + 2t_1t_2t_3 + 2t_1t_2t_4 + t_1t_3^2 \\ & + 2t_1t_3t_4 + t_1t_4^2 + t_2^2t_3 + t_2^2t_4 + t_2t_3^2 + 2t_2t_3t_4 \\ & + t_2t_4^2 + t_3^2t_4 + t_3t_4^2. \end{aligned} \tag{14.4}$$

The toric ideal $I_{\mathcal{A}_\lambda}$ is the ideal of algebraic relations among the 16 distinct monomials appearing in the expansion (14.4). An explicit quadratic Gröbner basis for this ideal is given in the proof of Proposition 14.4.

14.B. Permutation matrices

Let $\mathcal{A}$ be the set of all permutation matrices of size $p \times p$. In this example $d = p^2$, $n = p!$ and $dim(\mathcal{A}) = (p-1)^2 + 1$. The variables in the polynomial ring $k[\mathbf{x}]$ are indexed by the elements σ of the symmetric group S_p on $\{1,2,\ldots,p\}$, and $k[\mathbf{t}]$ is the ring of polynomial functions on a generic $p \times p$-matrix (t_{ij}). The toric map $\hat{\pi}$ in (4.2) takes the variable x_σ associated with the permutation σ to the monomial $t_{1\sigma(1)}t_{2\sigma(2)}\cdots t_{p\sigma(p)}$ which codes the permutation matrix $(\delta_{i,\sigma(i)})$. The polytope $Q = conv(\mathcal{A})$ is the famous *Birkhoff polytope* of doubly-stochastic matrices. It consists of all non-negative real $p \times p$-matrices with row sums and column sums equal to 1. Note that $dim(Q) = (p-1)^2$. Our objective is to study the toric ideals defined by the Birkhoff polytopes. Here are the first two cases.

Example 14.6. *(3 × 3-permutation matrices)*
For $p = 3$ the toric map $\hat{\pi} : k[\mathbf{x}] \to k[\mathbf{t}]$ is given by $x_{123} \mapsto t_{11}t_{22}t_{33}$, $x_{132} \mapsto t_{11}t_{23}t_{32}$, $x_{213} \mapsto t_{12}t_{21}t_{33}$, $x_{231} \mapsto t_{12}t_{23}t_{31}$, $x_{312} \mapsto t_{13}t_{21}t_{32}$ and $x_{321} \mapsto t_{13}t_{22}t_{31}$. The projective toric variety $Y_\mathcal{A}$ is a hypersurface of degree 3 in P^5. It is defined by the principal ideal

$$I_\mathcal{A} \quad = \quad \langle\, x_{123}x_{231}x_{312} \; - \; x_{132}x_{213}x_{321}\, \rangle.$$

Example 14.7. *(4 × 4-permutation matrices)*
The case $p = 4$ is more challenging: here $d = 16$, $n = 24$, and the projective toric variety $Y_\mathcal{A}$ has dimension 9 and degree 352. We order the variables in the usual lexicographic order on the symmetric group S_4, $x_{1234} \succ x_{1243} \succ x_{1324} \succ \cdots \succ x_{4321}$, and we define $\prec$ to be the induced *reverse lexicographic* term order on $k[\mathbf{x}]$. The reduced Gröbner basis $\mathcal{G}$ of $I_\mathcal{A}$ with respect to $\prec$ consists of 199 binomials: There are 18 quadrics such as

$$x_{1243}x_{2134} - x_{1234}x_{2143}, \quad x_{1324}x_{4231} - x_{1234}x_{4321}, \quad x_{1342}x_{3124} - x_{1324}x_{3142}, \ \ldots$$

The Gröbner basis $\mathcal{G}$ contains 176 cubics, including

$$x_{1234}x_{1342}x_{1423} - x_{1243}x_{1324}x_{1432},\ x_{1234}x_{1342}x_{4123} - x_{1243}x_{1324}x_{4132},\ \ldots\ldots,$$

and it contains precisely five quartics:

$$x_{1234}x_{1423}x_{3412}x_{4132} - x_{1432}^2x_{3124}x_{4213},\ x_{1342}x_{1423}x_{2431}x_{4132} - x_{1432}^2x_{2143}x_{4321},$$
$$x_{1342}x_{2314}x_{2431}x_{3241} - x_{1234}x_{2341}^2x_{3412},\ x_{2314}x_{3124}x_{3241}x_{4213} - x_{2143}x_{3214}^2x_{4321},$$
$$x_{1234}x_{2143}x_{3412}x_{4132} - x_{1432}x_{2134}x_{3142}x_{4213}.$$

In each displayed binomial the initial term comes first. All 199 initial terms are square-free. Thus $in_{\prec}(I_{\mathcal{A}})$ is a square-free monomial ideal. By our results of Chapter 8, it defines a regular triangulation of the Birkhoff polytope into 352 9-simplices of unit volume. ∎

Examples 14.6 and 14.7 are generalized in the following theorem.

Theorem 14.8. *Let $\succ$ be any of the $(p!)!$ graded reverse lexicographic term orders on $k[\mathbf{x}]$. Then the initial ideal $in_{\prec}(I_{\mathcal{A}})$ is generated by square-free monomials of degree $\leq p$.*

Proof: A key ingredient in our proof is the following well-known result from combinatorics (due to Birkhoff): *every non-negative integer $p \times p$-matrix with equal row and column sums can be written as a sum of permutation matrices.*

We fix any of the $(p!)!$ linear orders on the symmetric group S_p. Let $\mathcal{G}$ be the reduced Gröbner basis of $I_{\mathcal{A}}$ in the resulting reverse lexicographic term order on $k[\mathbf{x}]$. Consider any element $\mathbf{x}^{\mathbf{u}} - \mathbf{x}^{\mathbf{v}}$ of $\mathcal{G}$, where $\mathbf{x}^{\mathbf{u}}$ is the initial term. Let x_ρ denote the smallest variable which divides $\mathbf{x}^{\mathbf{v}}$. By the choice of reverse lexicographic order, it is smaller than any variable appearing in $\mathbf{x}^{\mathbf{u}}$. Writing $(u_\sigma)_{\sigma \in S_p}$ for the coordinates of the vector $\mathbf{u}$, we have

$$\hat{\pi}(\mathbf{x}^{\mathbf{v}}) \quad = \quad \hat{\pi}(\mathbf{x}^{\mathbf{u}}) \quad = \quad \prod_{\sigma \in S_p} \left(\hat{\pi}(x_\sigma)\right)^{u_\sigma} \quad = \quad \prod_{\sigma \in S_p} \prod_{i=1}^{p} t_{i\sigma(i)}^{u_\sigma}. \tag{14.5}$$

The monomial $\hat{\pi}(x_\rho) = t_{1\rho(1)} \cdots t_{p\rho(p)}$ divides (14.5). Therefore, for each index $i \in \{1, \ldots, p\}$ there exists a permutation σ such that $\sigma(i) = \rho(i)$ and $u_\sigma \geq 1$. Let $\mathbf{x}^{\mathbf{u}'}$ denote the product (without repetition) of the variables associated with these permutations. Then $\mathbf{x}^{\mathbf{u}'}$ is a square-free monomial of degree at most p. The condition $u_\sigma \geq 1$ guarantees that $\mathbf{x}^{\mathbf{u}'}$ divides $\mathbf{x}^{\mathbf{u}}$.

We claim that $\mathbf{x}^{\mathbf{u}'}$ lies in the initial ideal $in_{\prec}(I_{\mathcal{A}})$. If we divide $\hat{\pi}(\mathbf{x}^{\mathbf{u}'})$ by $\hat{\pi}(x_\rho)$ then we obtain a monomial $\prod_{i,j=1}^{p} t_{ij}^{c_{ij}}$ where (c_{ij}) is a non-negative integer $p \times p$-matrix with all row and column sums equal. By Birkhoff's Theorem, (c_{ij}) can be written as a sum of permutation matrices. Equivalently, the monomial $\prod_{i,j=1}^{p} t_{ij}^{c_{ij}}$ lies in the image of the map $\hat{\pi}$. Let $\mathbf{x}^{\mathbf{v}'}$ be any of its preimages. By construction, the two monomials $\mathbf{x}^{\mathbf{u}'}$ and $x_\rho \cdot \mathbf{x}^{\mathbf{v}'}$ have the same image under $\hat{\pi}$. Therefore $\mathbf{x}^{\mathbf{u}'} - x_\rho \cdot \mathbf{x}^{\mathbf{v}'}$ lies in $I_{\mathcal{A}}$. Here $\mathbf{x}^{\mathbf{u}'}$ is the initial term since x_ρ is smaller than any of the variables in $\mathbf{x}^{\mathbf{u}'}$. This proves our claim.

Now, $\mathbf{x}^{\mathbf{u}}$ is both a minimal generator of $in_{\prec}(I_{\mathcal{A}})$ and a multiple of $\mathbf{x}^{\mathbf{u}'}$. This implies $\mathbf{x}^{\mathbf{u}} = \mathbf{x}^{\mathbf{u}'}$. Thus $in_{\prec}(I_{\mathcal{A}})$ is generated by square-free monomials of degree $\leq p$. ∎

Corollary 14.9. (Stanley 1980) *Every reverse lexicographic triangulation of the Birkhoff polytope of doubly-stochastic matrices is unimodular.*

Proof: By Corollary 8.9 and Theorem 14.8. ■

Stanley (1980) calls a lattice polytope $Q = conv(\mathcal{A})$ *compressed* if each of its reverse lexicographic triangulations is unimodular. Equivalently, we may call a homogeneous toric ideal $I_{\mathcal{A}}$ *compressed* if every reverse lexicographic initial ideal of $I_{\mathcal{A}}$ is square-free. Being compressed is a strictly weaker property than being unimodular, since there may be other initial ideals which are not square-free. The simplest example of a compressed polytope which is not unimodular is the regular 3-dimensional cube.

It is easy to see that the Birkhoff polytope is not unimodular for $p \geq 4$. Consider the binomial $x_{1234}x_{1423}x_{3412}x_{4132} - \underline{x_{1432}^2 x_{3124} x_{4213}}$ which appears in the reduced Gröbner basis of Example 14.7. By Corollary 7.9, there exists a term order $\prec$ on $k[\mathbf{x}]$ such that this binomial appears in the reduced Gröbner basis with the underlined term as its initial term. The monomial ideal $in_\prec(I_{\mathcal{A}})$ is not square-free, and hence the set $\mathcal{A}$ of 4×4-permutation matrices is not unimodular (by Remark 8.10). This example lifts to $p \geq 5$.

The reverse lexicographic Gröbner bases of a compressed toric ideal have the pleasant property that they can be used for normal form reduction without knowing them explicitly. For instance, if $p \geq 5$ then it is a rather hopeless enterprise to explicitly compute any reduced Gröbner basis for the toric ideal of the Birkhoff polytope. Nevertheless we can use the Gröbner bases of Theorem 14.8 for normal form reduction in an implicit manner. Fix the usual lexicographic order on the symmetric group S_p with the identity $123\ldots p$ the highest element and its reverse $p(p-1)\ldots 21$ the lowest element. Let $\prec$ denote the resulting reverse lexicographic term order on $k[\mathbf{x}]$. The following procedure computes the normal form of a monomial modulo the "Birkhoff ideal" $I_{\mathcal{A}}$.

Algorithm 14.10. *(Normal form modulo the toric ideal of the Birkhoff polytope)*
Input: A monomial $\mathbf{x}^{\mathbf{u}}$ in the variables x_σ, $\sigma \in S_p$.
Output: The normal form of $\mathbf{x}^{\mathbf{u}}$ modulo the ideal $I_{\mathcal{A}}$ with respect to the term order $\prec$.

1. Let r be the degree of $\mathbf{x}^{\mathbf{u}}$.
2. Write $\mathbf{x}^{\mathbf{u}}$ as an $r \times p$-tableau $\mathbf{U}$ with entries in $\{1, 2, \ldots, p\}$ whose rows are the permutations corresponding to the variables in $\mathbf{x}^{\mathbf{u}}$, with repetition.
3. While $r > 0$ do
   - 3.1. let σ be the smallest permutation obtainable by taking one entry from each column of $\mathbf{U}$. Delete these p entries from $\mathbf{U}$ while restoring the rectangular shape (as in the example below), and let $\mathbf{U}'$ be the resulting $(r-1) \times p$-tableau
   - 3.2. Output the variable x_σ. Set $r := r - 1$ and $\mathbf{U} := \mathbf{U}'$.
4. The normal form is the product of all variables which have been output along the way.

The correctness of this algorithm follows from the proof of Theorem 14.8. Here is an example how it works. Let $p = 5$ and consider the monomial

$$\mathbf{x}^{\mathbf{u}} = x_{12435} \cdot x_{25134} \cdot x_{31452} \cdot x_{43125} \cdot x_{53241}$$

The sequence of tableaux generated by Algorithm 14.10 equals

$$\begin{bmatrix} 1 & 2 & 4 & 3 & 5 \\ 2 & 5 & 1 & 3 & 4 \\ 3 & 1 & \mathbf{4} & 5 & 2 \\ 4 & \mathbf{3} & 1 & \mathbf{2} & 5 \\ \mathbf{5} & 3 & 2 & 4 & \mathbf{1} \end{bmatrix} \rightarrow \begin{bmatrix} 1 & 2 & 4 & 3 & 5 \\ 2 & \mathbf{5} & 1 & \mathbf{3} & 4 \\ 3 & 1 & \mathbf{1} & 5 & \mathbf{2} \\ \mathbf{4} & 3 & 2 & 4 & 5 \end{bmatrix} \rightarrow \begin{bmatrix} 1 & \mathbf{2} & 4 & 3 & 5 \\ 2 & 1 & \mathbf{1} & \mathbf{5} & \mathbf{4} \\ \mathbf{3} & 3 & 2 & 4 & 5 \end{bmatrix}$$

$$\rightarrow \begin{bmatrix} 1 & \mathbf{1} & \mathbf{4} & \mathbf{3} & \mathbf{5} \\ \mathbf{2} & 3 & 2 & 4 & 5 \end{bmatrix} \rightarrow \begin{bmatrix} 1 & 3 & 2 & 4 & 5 \end{bmatrix}$$

We conclude that the normal form of $\mathbf{x}^{\mathbf{u}}$ equals

$$\mathbf{x}^{\mathbf{u}'} \quad = \quad x_{53421} \cdot x_{45132} \cdot x_{32154} \cdot x_{21435} \cdot x_{13245}.$$

The study of the toric ideal defined by the permutation matrices is motivated by a sampling problem from statistics. In fact, it may be added that it was this very problem which got the project (Diaconis & Sturmfels 1993) started. Consider a group of r voters electing a president from a list of p candidates. Each of the r voters ranks the p candidates in order of preference. This collection of r permutations is the complete *election data*. To compress this mass of information we form a $p \times p$-matrix as follows: the entry in position (i,j) is the number of people who rank candidate i in position j. This matrix is the *first-order summary*. The sampling problem is to choose at random from the set of all election data with a fixed first-order summary. When modeling this problem as in Chapter 5, one realizes that the set $\mathcal{A}$ is precisely the set of $p \times p$-permutation matrices. The election data is a monomial $\mathbf{x}^{\mathbf{u}}$ in $p!$ variables, and its first order-summary is a monomial $\hat{\pi}(\mathbf{x}^{\mathbf{u}}) = \mathbf{t}^{\pi(\mathbf{u})}$ in p^2 variables. Theorem 14.8 implies that any two election data with the same first-order summary can be connected by a sequence of local moves involving at most p voters.

14.C. Three-dimensional matrices

Our next example comes from a direct generalization of the sampling problem in Example 5.1. In (5.4) a group of people was classified according to two features (hair color and eye color). It is natural to consider three or more features and study their interrelations. This leads us to introduce three-dimensional matrices (of format $r \times s \times t$). The case of $3 \times 3 \times 3$-matrices appeared in Exercise (7) of Chapter 5. The corresponding integer programming problem is known as the *three-dimensional transportation problem*; see e.g. (Vlach 1986).

Fix integers $r \leq s \leq t$. Let $n = rst$, $d = rs + rt + st$, and identify $\mathbf{Z}^d$ with the direct sum of matrices spaces $\mathbf{Z}^{r\times s} \oplus \mathbf{Z}^{r\times t} \oplus \mathbf{Z}^{s\times t}$. We denote the standard basis vectors in the three components as $\mathbf{e}_{ij}$, $\mathbf{e}'_{ik}$ and $\mathbf{e}''_{jk}$ respectively. Our configuration in this subsection is

$$\mathcal{A} \quad = \quad \left\{ \mathbf{e}_{ij} \oplus \mathbf{e}'_{ik} \oplus \mathbf{e}''_{jk} \;:\; i = 1,\ldots,r,\; j = 1,\ldots,s,\; k = 1,\ldots,t \right\}. \tag{14.6}$$

The resulting ring map is

$$\hat{\pi} \,:\, k\big[x_{ijk}\big] \;\rightarrow\; k\big[u_{ij}, v_{ik}, w_{jk}\big]\,, \quad x_{ijk} \;\mapsto\; u_{ij}\cdot v_{ik}\cdot w_{jk}, \tag{14.7}$$

where the indices i, j, k run as in (14.6). At first glance the projective toric variety $Y_{\mathcal{A}}$ may appear similar to a Segre variety. But this a deception. The kernel of (14.7) is much more complicated than the Segre ideal (5.7). Formula (4.3) translates into the following:

$$I_{\mathcal{A}} = \Big\langle \prod_{ijk} x_{ijk}^{a_{ijk}} - \prod_{ijk} x_{ijk}^{b_{ijk}} \; : \; \sum_{i=1}^{r} a_{ijk} = \sum_{i=1}^{r} b_{ijk} \text{ for all } j,k, \\ \sum_{j=1}^{s} a_{ijk} = \sum_{j=1}^{s} b_{ijk} \text{ for all } i,k, \; \sum_{k=1}^{t} a_{ijk} = \sum_{k=1}^{t} b_{ijk} \text{ for all } i,j \Big\rangle. \tag{14.8}$$

To emphasize the combinatorial difficulty of the configuration $\mathcal{A}$ in (14.6), we remark that it is an unsolved problem to describe the facets of *conv*$(\mathcal{A})$. Partial results on this problem are given in (Vlach 1986).

Our first example is the case $r = s = t = 2$. Here the toric ideal is principal:

$$I_{\mathcal{A}} \quad = \quad \langle\, x_{111}x_{122}x_{212}x_{221} \; - \; x_{112}x_{121}x_{211}x_{222}\,\rangle.$$

It turns out that the case $r = 2$ and s, t arbitrary is still easy, in the sense that it can be reduced to Proposition 5.4. We write $\mathcal{A}_{st} := \{\, \mathbf{e}_i \oplus \mathbf{e}'_j \,\}$ for the *Segre configuration* in Example 5.1. Thus $I_{\mathcal{A}_{st}}$ is the ideal of 2×2-minors of an $s\times t$-matrix of indeterminates.

Proposition 14.11. *If $r = 2$ then the configuration $\mathcal{A}$ in (14.6) isomorphic to the Lawrence lifting $\Lambda(\mathcal{A}_{st})$ of the Segre configuration $\mathcal{A}_{st}$. In particular, $\mathcal{A}$ is unimodular.*

Proof: To exhibit the isomorphism we rename the variables $x_{jk} := x_{1jk}$ and $y_{jk} := x_{2jk}$. Consider the binomials in (14.8) which have no monomial factors. The first set of conditions translates into $a_{1jk} = b_{2jk}$ and $a_{2jk} = b_{1jk}$ for $j = 1, \ldots, s$ and $k = 1, \ldots, t$. Using these we replace all b-variables. The second and third set of conditions in (14.8) now equals

$$\sum_{j=1}^{s} (a_{1jk} - a_{2jk}) = 0 \;\; \text{for all } k = 1, \ldots, t \quad \text{and}$$

$$\sum_{k=1}^{t} (a_{1jk} - a_{2jk}) = 0 \;\; \text{for all } j = 1, \ldots, s.$$

This says that the $s\times t$-matrix $\mathbf{c} = (c_{jk}) := (a_{1jk} - a_{2jk})$ has zero row and column sums. Our assumption (no monomial factor) implies that $a_{1jk} = 0$ or $a_{2jk} = 0$. Therefore $\mathbf{c}^{+} = (a_{1jk})$ and $\mathbf{c}^{-} = (a_{2jk})$. With all this new notation equation (14.8) becomes

$$I_{\mathcal{A}} \quad = \quad \langle\, \mathbf{x}^{\mathbf{c}^{+}}\mathbf{y}^{\mathbf{c}^{-}} - \mathbf{x}^{\mathbf{c}^{-}}\mathbf{y}^{\mathbf{c}^{+}} \;:\; \mathbf{c} \in \mathbf{Z}^{s\times t} \text{ has zero row and column sums }\rangle.$$

Formula (7.1) shows that $I_{\mathcal{A}}$ is the toric ideal of the Lawrence lifting of $\mathcal{A}_{st}$. The fact that $\mathcal{A}$ and $\Lambda(\mathcal{A}_{st})$ have the same toric ideal means that these two vector configurations are isomorphic (i.e., there is an isomorphism between their ambient d-dimensional lattices which carries one into the other). Finally, $\mathcal{A}$ is unimodular because the configuration $\mathcal{A}_{rs}$ is unimodular and the operator $\Lambda(\,\cdot\,)$ clearly preserves this property. ∎

For $r = 2$ and s, t arbitrary, the Graver basis of $\mathcal{A}$ is given by the following corollary.

Corollary 14.12. *The Graver basis $Gr_{\Lambda(\mathcal{A}_{rs})}$ consists of all binomials*

$$x_{j_1k_1}x_{j_2k_2}\cdots x_{j_sk_s}y_{j_2k_1}y_{j_3k_2}\cdots y_{j_1k_s} - x_{j_2k_1}x_{j_3k_2}\cdots x_{j_1k_s}y_{j_1k_1}y_{j_2k_2}\cdots y_{j_sk_s}, \quad (14.9)$$

where $(j_1,k_1),(k_1,j_2),\ldots,(j_s,k_s),(k_s,j_1)$ is a circuit in the complete bipartite graph K_{rs}.

Proof: Erase all the y-variables in the binomials (14.9). In view of equation (7.2), it suffices to show that these binomials form the Graver basis of the determinantal ideal $I_{\mathcal{A}_{rs}}$. (Perhaps you have done this already when you solved Exercise (1) of Chapter 5 ?) Since $\mathcal{A}_{rs}$ is unimodular, we have $\mathcal{C}_{\mathcal{A}_{rs}} = Gr_{\mathcal{A}_{rs}}$, by Proposition 8.11. But the circuits of $\mathcal{A}_{rs}$ are precisely the circuits of the complete bipartite graph K_{rs} (see Lemma 9.8 for a more general statement). ∎

The next case is $r=s=t=3$. The toric ideal $I_{\mathcal{A}}$ of the $3\times 3\times 3$-transportation problem lives in a polynomial ring in 27 variables. The projective variety $Y_{\mathcal{A}}$ has dimension 9. By direct computation we found that the set of circuits still coincides with the Graver basis. This is surprising since the configuration $\mathcal{A}$ is no longer unimodular. The non-unimodularity is seen from the appearance of squares in the circuits of types (f) and (g).

Theorem 14.13. *For $r = s = t = 3$ the Graver basis $Gr_{\mathcal{A}}$ equals the set $\mathcal{C}_{\mathcal{A}}$ of circuits. There are precisely 795 circuits; they are grouped into seven symmetry classes as follows:*

(a) 27 circuits of degree 4 such as

$$x_{123}\,x_{132}\,x_{322}\,x_{333} - x_{122}\,x_{133}\,x_{323}\,x_{332},$$

(b) 54 circuits of degree 6 such as

$$x_{111}\,x_{122}\,x_{212}\,x_{223}\,x_{313}\,x_{321} - x_{112}\,x_{121}\,x_{213}\,x_{222}\,x_{311}\,x_{323}$$

(c) 108 circuits of degree 7 such as

$$x_{111}\,x_{123}\,x_{132}\,x_{222}\,x_{231}\,x_{313}\,x_{321} - x_{113}\,x_{122}\,x_{131}\,x_{221}\,x_{232}\,x_{311}\,x_{323}$$

(d) 216 circuits of degree 9 such as

$$x_{112}\,x_{121}\,x_{133}\,x_{222}\,x_{231}\,x_{311}\,x_{323}\,x_{332}^2 \\ - x_{111}\,x_{123}\,x_{132}\,x_{221}\,x_{232}\,x_{312}\,x_{322}\,x_{331}\,x_{333}$$

(e) 12 circuits of degree 9 such as

$$x_{112}\,x_{123}\,x_{131}\,x_{213}\,x_{221}\,x_{232}\,x_{311}\,x_{322}\,x_{333} \\ - x_{113}\,x_{121}\,x_{132}\,x_{211}\,x_{222}\,x_{233}\,x_{312}\,x_{323}\,x_{331}$$

(f) 162 circuits of degree 10 such as

$$x_{111}\,x_{123}\,x_{132}\,x_{213}^2\,x_{221}\,x_{231}\,x_{311}\,x_{322}\,x_{333} \\ - x_{113}\,x_{122}\,x_{131}\,x_{211}^2\,x_{223}\,x_{233}\,x_{313}\,x_{321}\,x_{332}$$

(g) 216 circuits of degree 12 such as

$$x_{111}\,x_{123}^2\,x_{132}^2\,x_{213}\,x_{222}^2x_{231}\,x_{312}\,x_{321}\,x_{333}$$
$$-\quad x_{113}\,x_{122}^2\,x_{131}\,x_{133}\,x_{212}\,x_{221}\,x_{223}\,x_{232}\,x_{311}\,x_{323}\,x_{332}$$

Moreover, the 81 *quartic and sextic binomials of types* (a) *and* (b) *minimally generate* $I_{\mathcal{A}}$.

Proof: The 795 circuits were computed by a brute-force approach based on formula (1.1). Our proof that they constitute the Graver basis will be explained in detail. We introduce 27 new variables $y_{111}, y_{112}, \ldots, y_{333}$. We homogenize each of the above 795 circuits as follows: if the variable x_{ijk} appears in one of the two terms then multiply the other term by y_{ijk}. For example, the homogenization of the quartic in (a) is

$$x_{123}\,x_{132}\,x_{322}\,x_{333}\,y_{122}\,y_{133}\,y_{323}\,y_{332} \quad - \quad x_{122}\,x_{133}\,x_{323}\,x_{332}\,y_{123}\,y_{132}\,y_{322}\,y_{333}.$$

The resulting 795 binomials are the circuits of $\Lambda(\mathcal{A})$, by part (b) of Exercise (3) below. In view of equation (7.2), it suffices to show that these 795 binomials form the Graver basis for $I_{\Lambda(\mathcal{A})}$. In view of Theorem 7.1, it suffices to simply show that they generate $I_{\Lambda(\mathcal{A})}$.

Let J be the ideal generated by the given 795 circuits of $\Lambda(\mathcal{A})$. We claim that $J = I_{\Lambda(\mathcal{A})}$. Let X denote the product of all 54 variables. We shall make use of Lemma 12.2, which states that $(J : X^\infty) = I_{\Lambda(\mathcal{A})}$. Our claim is therefore reduced to the assertion

$$(J : x_{ijk}) \;=\; J \qquad \text{and} \qquad (J : y_{ijk}) \;=\; J \qquad \text{for all } 1 \leq i,j,k \leq 3. \qquad (14.10)$$

Consider the symmetry group of the configuration $\Lambda(\mathcal{A})$. Clearly, it acts transitively on the 54 variables, and it leaves the ideal J invariant. Therefore it suffices to check the condition (14.10) for one variable only, say x_{111}. We shall apply Lemma 12.1. Fix any reverse lexicographic term order on $k[\mathbf{x},\mathbf{y}]$ such that x_{111} is the cheapest variable. A single application of Buchberger's criterion reveals that the given 795 binomials in 54 variables are already a reduced Gröbner basis in this order. (This involves the computation of up to $\binom{795}{2} = 315,615$ S-pairs.) By Lemma 12.1, this implies $(J : x_{111}) = J$, and we are done.

Our last claim that the binomials in (a) and (b) form a minimal generating set was verified by a brute force computation in MACAULAY. We have proved Theorem 14.13. ∎

A binomial in one of the ideals $I_{\mathcal{A}}$ is said to have *format* (r,s,t) if its variables x_{ijk} involve r distinct indices i, s distinct indices j, and t distinct indices k. For instance, among the circuits in Theorem 14.13, the binomial in (a) has format $2\times 2\times 2$, the binomial in (b) has format $3\times 2\times 3$, and all others have format $3\times 3\times 3$. Hence we observe that the ideal for $3\times 3\times 3$-matrices is generated by binomials of strictly smaller format. This raises the question whether there exist R, S and T such that, for all $r \geq R, s \geq S, t \geq T$, our ideal is generated by binomials of format smaller than $R \times S \times T$. The answer is "no":

Proposition 14.14. *For every positive even integer p there exists a minimal generator for the toric ideal (14.8) which has degree $2p$ and format $2p \times (p+1) \times (p+1)$.*

Proof: For $p > 0$ even we consider the following Laurant monomial

$$\prod_{i=1}^{p} \left(\frac{x_{i,i+1,i} \cdot x_{i,i,i+1}}{x_{i,i,i} \cdot x_{i,i+1,i+1}} \right)^{(-1)^i} \cdot \frac{x_{p+1,1,3} \cdot x_{p+1,2,1}}{x_{p+1,1,1} \cdot x_{p+1,2,3}} \cdot \prod_{\substack{i=2 \\ i \text{ even}}}^{p-1} \left(\frac{x_{p+1,i-1,i} \cdot x_{p+1,i+2,i+1}}{x_{p+1,i-1,i+1} \cdot x_{p+1,i+2,i}} \right)$$

$$\times \prod_{\substack{i=2 \\ i \text{ odd}}}^{p-1} \left(\frac{x_{p+i,i,i+2} \cdot x_{p+i,i+1,i-1}}{x_{p+i,i,i-1} \cdot x_{p+i,i+1,i+2}} \right) \cdot \frac{x_{2p,p-1,p} \cdot x_{2p,p+1,p+1}}{x_{2p,p-1,p+1} \cdot x_{2p,p+1,p}}$$

We write $\mathbf{x^u}$ for the numerator and $\mathbf{x^v}$ for the denominator of this expression. Then $\mathbf{x^u} - \mathbf{x^v}$ is a binomial of degree $2p$ and format $2p \times (p+1) \times (p+1)$. It can be checked that $\mathbf{x^u} - \mathbf{x^v}$ lies in the kernel of (14.7). The general pattern is best understood by looking at the first two instances. For $p = 2$ our binomial equals

$$x_{111}\, x_{122}\, x_{232}\, x_{223}\, x_{313}\, x_{321}\, x_{412}\, x_{433} \;-\; x_{121}\, x_{112}\, x_{222}\, x_{233}\, x_{311}\, x_{323}\, x_{413}\, x_{432}$$

and for $p = 4$ it equals

$$x_{111}\, x_{122}\, x_{232}\, x_{223}\, x_{333}\, x_{344}\, x_{454}\, x_{445}\, x_{521}\, x_{512}\, x_{543}\, x_{735}\, x_{742}\, x_{834}\, x_{855}$$
$$-\; x_{121}\, x_{112}\, x_{222}\, x_{233}\, x_{343}\, x_{334}\, x_{444}\, x_{455}\, x_{511}\, x_{523}\, x_{542}\, x_{732}\, x_{745}\, x_{835}\, x_{854}.$$

We claim that $\mathbf{x^u} - \mathbf{x^v}$ is a minimal generator of the toric ideal. In view of Corollary 12.13, it suffices to show that $\mathbf{u}$ and $\mathbf{v}$ are the only two elements in their common fiber. Suppose that $\mathbf{w}$ is any other $2p \times (p+1) \times (p+1)$-matrix in the same fiber. For fixed second and third index jk there is at most one variable $x_{\cdot jk}$ appearing in $\mathbf{x^w}$. This holds because each image variable w_{jk} appears at most linearly in $\hat{\pi}(\mathbf{x^w}) = \hat{\pi}(\mathbf{x^v}) = \hat{\pi}(\mathbf{x^u})$. Now, for each fixed first index i our binomial $\mathbf{x^u} - \mathbf{x^v}$ looks like a ordinary 2×2-determinant $x_{i\cdot\cdot} x_{i\cdot\cdot} - x_{i\cdot\cdot} x_{i\cdot\cdot}$. One the these two terms must appear as a factor of $\mathbf{x^w}$. These two requirements together imply that $\mathbf{x^w} = \mathbf{x^u}$ or $\mathbf{x^w} = \mathbf{x^v}$. ∎

Exercises:

(1) How many of the 199 Gröbner basis elements in Example 14.7 are needed to minimally generate this toric ideal ? Are the quadrics and cubics sufficient ?

(2) Estimate the size of a reverse lexicographic Gröbner basis for the toric ideal defined by the 5×5-permutation matrices. Determine the exact number of Gröbner basis elements of degree two and three.

(3) Let $\mathcal{A} = \{\mathbf{a}_1, \ldots, \mathbf{a}_n\}$ be any configuration in $\mathbf{Z}^d$. Prove the following two statements:
   (a) The set of circuits $\mathcal{C}_{\mathcal{A}}$ generates the lattice $ker_{\mathbf{Z}}(\mathcal{A})$ (as an abelian group).
   (b) Let $\Lambda(\mathcal{A}) \subset \mathbf{Z}^{d+n}$ be the Lawrence lifting. Then $\mathcal{C}_{\Lambda(\mathcal{A})} = \{(\mathbf{u}, -\mathbf{u}) : \mathbf{u} \in \mathcal{C}_{\mathcal{A}}\}$.

(4) Consider the cubic Veronese embedding of P^3 into P^{19}. The quadratic Gröbner basis in Theorem 14.2 defines a regular triangulation consisting of 27 tetrahedra. List them.

Notes:

Eisenbud, Reeves and Totaro (1994) have given a quadratic Gröbner basis for the Veronese ideal, but theirs is not square-free. The square-free quadratic Gröbner basis for the Veronese ideal appearing in Theorem 14.2 was first derived from the corresponding triangulation, which is the special regular triangulation appearing in Section III.2B of (Kempf, Knudsen, Mumford & Saint-Donat 1973). The examples and results in subsections 14.B and 14.C are taken from (Diaconis & Sturmfels 1993).

Bibliography

Adams, W.W., Loustaunau, P. (1994). *An Introduction to Gröbner Bases,* American Mathematical Society, Graduate Studies in Math. Vol. III.

Arnold, V.I. (1989). A-graded algebras and continued fractions, *Communications in Pure and Applied Math,* **42**, 993–1000.

Aspinwall, P., Greene, B., Morrison, D. (1993). The monomial-divisor mirror map, *Internat. Math. Res. Notices* **12**, 319–337.

Avis, D., Fukuda, K. (1992). A pivoting algorithm for convex hulls and vertex enumeration of arrangements and polyhedra, *Discrete Comput. Geom.* **8**, 295–313.

Batyrev, V. (1993). Quantum cohomology rings of toric manifolds, Journees de Geometrie Algebrique d'Orsay 1992, *Asterisque,* **218**, 9–34.

Bayer, D. (1982). *The Division Algorithm and the Hilbert Scheme,* Ph. D. Dissertation, Harvard University.

Bayer, D., Morrison, I. (1988). Gröbner bases and geometric invariant theory I, *J. Symbolic Computation* **6**, 209-217.

Bayer, D., Stillman, M. (1987a). A criterion for detecting m-regularity, *Inventiones Math.* **87**, 1–11.

Bayer, D., Stillman, M. (1987b). Macaulay: a computer algebra system for algebraic geometry, available by anonymous ftp from `zariski.harvard.edu`.

Becker, T., Kredel, H., Weispfenning, V. (1993). *Gröbner Bases: A Computational Approach to Commutative Algebra,* Graduate Texts in Mathematics, Springer, New York.

Billera, L.J., Filliman, P., Sturmfels, B. (1990). Constructions and complexity of secondary polytopes, *Advances in Mathematics* **83**, 155–179.

Björner, A., Las Vergnas, M., Sturmfels, B., White, N., Ziegler, G. (1993). *Oriented Matroids,* Cambridge University Press.

Bouvier, C., Gonzalez-Sprinberg, G. (1992). G-desingularisations de variétes toriques, *C.R. Acad. Sci. Paris, Ser. I* **315**, 817–820.

Bresinsky, H. (1988). Binomial generating sets for monomial curves, with applications in $\mathbf{A}^4$, *Rend. Sem. Mat. Univ. Politec. Torino* **46**, 353–370.

Bruns, W., Herzog, J. (1993). *Cohen-Macaulay Rings,* Cambridge University Press.

Buchberger, B. (1985). Gröbner bases – an algorithmic method in polynomial ideal theory, Chapter 6 in N.K. Bose (ed.) : *Multidimensional Systems Theory,* D. Reidel Publ.

Campillo, A., Marijuan, C. (1991). Higher relations for a numerical semigroup, *Sém. Théorie Nombres Bordeaux* **3**, 249–260.

Campillo, A., Pison, P. (1993). L'idéal d'un semi-group de type fini, *Comptes Rendues Acad. Sci. Paris*, Série I, **316**, 1303–1306.

Collart, S., Kalkbrener, M., Mall, D. (1996). Converting Bases with the Gröbner Walk, *Journal of Symbolic Computation*, to appear.

Conti, P., Traverso, C. (1991). Buchberger algorithm and integer programming, Proceedings AAECC-9 (New Orleans), Springer LNCS **539**, pp. 130-139.

Cox, D., Little, J., O'Shea, D. (1992). *Ideals, Varieties and Algorithms*, Undergraduate Texts in Mathematics, Springer, New York.

Dabrowski, R. (1996). On normality of the closure of a generic torus orbit in G/P, *Pacific J. Math.*, **172**, 321–330.

De Loera, J. (1995a). Non-regular triangulations of products of simplices, *Discrete and Computational Geometry*, to appear.

De Loera, J. (1995b). *Triangulations of Polytopes and Computational Algebra*, Ph. D. dissertation, Cornell University.

De Loera, J., Sturmfels, B., Thomas, R. (1995). Gröbner bases and triangulations of the second hypersimplex, *Combinatorica*, **15**, 409–424.

Diaconis, P., Sturmfels, B. (1993). Algebraic algorithms for sampling from conditional distributions, *Annals of Statistics*, to appear.

Diaconis, P., Graham, R., Sturmfels, B. (1995). Primitive partition identities, in: *Paul Erdös is 80. Vol II*, Janos Bolyai Society, Budapest, pp. 1–20.

Di Biase, F., Urbanke, R. (1995). An algorithm to compute the kernel of certain polynomial ring homomorphisms, *Experimental Mathematics*, **4**, 227–234.

Edelsbrunner, H. (1987). Algorithms in Combinatorial Geometry. Springer, New York.

Eisenbud, D. (1995). *Introduction to Commutative Algebra with a View Towards Algebraic Geometry*, Graduate Texts in Mathematics, Springer, New York.

Eisenbud, D., Reeves, A., Totaro, B. (1994). Initial ideals, Veronese subrings, and rates of algebras. *Advances in Mathematics* **109**, 168–187.

Eisenbud, D., Sturmfels, B. (1996). Binomial ideals, *Duke Mathematical Journal*, **84**, 1–45.

Ewald, G., Schmeinck, A. (1993). Representation of the Hirzebruch-Kleinschmidt varieties by quadrics, *Beiträge zur Algebra und Geometrie*, **34**, 151–156.

Faugère, J.C., Gianni, P., Lazard, D., Mora, T. (1992). Efficient computation of zero-dimensional Gröbner bases by change of ordering. *J. Symbolic Computation*, **13**, 117–131.

Fulton, W. (1993). *Introduction to Toric Varieties*, Princeton University Press.

Gel'fand, I., Kapranov, M., Zelevinsky, A. (1994). *Discriminants, Resultants and Multi-Dimensional Determinants*, Birkhäuser, Boston.

Göbel, M. (1995). Computing bases for rings of permutation-invariant polynomials, *Journal of Symbolic Computation*, **19**, 285–291.

Graver, J. (1975). On the foundations of linear and integer programming I, *Mathematical Programming* **8**, 207–226.

Gritzmann, P., Sturmfels, B. (1993). Minkowski addition of polytopes: Computational complexity and applications to Gröbner bases, *SIAM J. Discrete Math.* **6**, 246–269.

Grünbaum, B. (1967). *Convex Polytopes*, Wiley Interscience, London.

Hartshorne, R. (1977). *Algebraic Geometry*, Springer Verlag, New York.

Herzog, J. (1970). Generators and relations of semigroups and semigroup rings, *Manuscripta Math.* **3**, 175–193.

Hosten, S., Sturmfels, B. (1995). GRIN: An implementation of Gröbner bases for integer programming, in "Integer Programming and Combinatorial Optimization" (E. Balas, J. Clausen, eds.), Springer Lecture Notes in Computer Science, **920**, pp. 267–276.

Kalkbrener, M., Sturmfels, B. (1995). Initial complexes of prime ideals, *Advances in Mathematics*, **116**, 365–376.

Kapranov, M., Sturmfels, B., Zelevinsky, A. (1992). Chow polytopes and general resultants, *Duke Mathematical Journal* **67**, 189–218.

Kapur, D., Madlener, K. (1989). A completion procedure for computing a canonical basis of a k-subalgebra. Proceedings of *Computers and Mathematics 89* (eds. Kaltofen and Watt), MIT, Cambridge, June 1989.

Kempf, G., Knudsen, F., Mumford, D., and Saint-Donat, B. (1973). *Toroidal Embeddings*, Springer Lecture Notes in Mathematics **339**.

Koelman, R. (1993a). Generators for the ideal of a projectively embedded toric surface, *Tôhoku Math. Journal* **45**, 385–392.

Koelman, R. (1993b). A criterion for the ideal of a projectively embedded toric surface to be generated by quadrics, *Beiträge zur Algebra und Geometrie*, **34**, 57–62.

Korkina, E., Post, G., Roelofs, M. (1995). Classification of generalized A-graded algebras with 3 generators, *Bulletin des Sciences Mathématiques*, **119**, 267–287.

Lee, C. (1991). Regular triangulations of convex polytopes, in *Applied Geometry and Discrete Mathematics - The Victor Klee Festschrift*, (P. Gritzmann and B. Sturmfels, eds.), American Math. Soc, DIMACS Series **4**, Providence, R.I., pp. 443–456.

Lovász, L., Plummer, M.D. (1986). *Matching Theory*, North-Holland, New York, (also in *Annals in Discrete Mathematics* **29**).

Mall, D. (1995). Combinatorics of Polynomial Ideals and Gröbner Bases, Habilitationsschrift, ETH Zürich.

Miller, J.L. (1996). Analogs of Gröbner bases in polynomial rings over a ring, *Journal of Symbolic Computation* **21**, 139–153.

Möller, H.M., Mora, T. (1984). Upper and lower bounds for the degree of Gröbner bases, EUROSAM '84, *Springer Lecture Notes in Computer Science* **174**, 172–183.

Mora, T., Robbiano, L. (1988). The Gröbner fan of an ideal, *J. Symbolic Computation* **6** 183-208.

Mount, J. (1995). *Application of Convex Sampling to Optimization and Contingency Table Generation/Counting*, Ph.D. Dissertation, Department of Computer Science, Carnegie Mellon University.

Oda, T. (1988). *Convex Bodies and Algebraic Geometry: an Introduction to the Theory of Toric Varieties.* Springer Verlag, New York.

Ollivier, F. (1991). Canonical bases: relations with standard bases, finiteness conditions, and applications to tame automorphisms, in *Effective Methods in Algebraic Geometry* (T. Mora, C. Traverso, eds). Birkhäuser, Boston, pp. 379–400.

Ostrowski, A. (1921). Über die Bedeutung der Theorie der konvexen Polyeder für die formale Algebra, *Jahresberichte Deutsche Math. Verein.* **30**, 98–99

Pottier, L. (1994). Gröbner bases of toric ideals, Rapport de recherche 2224, INRIA Sophia Antipolis, Manuscript available from `http://www.inria.fr/safir/SAFIR/Loic.html`.

Reeves, A., Sturmfels, B. (1993). A note on polynomial reduction, *Journal of Symbolic Computation* **11**, 273–277.

Rippel, C. (1994). *Generic Initial Ideal Theory for Coordinate Rings of Flag Varieties*, Ph.D. Dissertation, UCLA.

Robbiano, L., Sweedler, M. (1990). Subalgebra bases, in *Springer Lecture Notes in Mathematics* **1430**, pp. 61-87.

Schrijver, A. (1986). *Theory of Linear and Integer Programming,* Wiley Interscience, Chichester 1986.

Schwartz, N. (1988). Stability of Gröbner bases, *J. Pure Appl. Algebra* **53**, 171–186.

Shafarevich, I.R. (1977). *Basic Algebraic Geometry*, Study Edition, Springer Verlag, Berlin.

Sinclair, A. (1993). *Algorithms for Random Generation and Counting: A Markov Chain Approach*, Birkhäuser, Boston, 1993.

Stanley, R. (1980). Decomposition of rational convex polytopes, *Annals of Discrete Math.* **6**, 333-342.

Stanley, R. (1986). *Enumerative Combinatorics.* Wadsworth & Brooks, Monterey, CA.

Stanley, R. (1987). Generalized H-vectors, intersection cohomology of toric varieties, and related results. in *Commutative Algebra and Combinatorics, Advanced Studies in Pure Math.* **11**, 187–213.

Sturmfels, B. (1991). Gröbner bases of toric varieties, *Tôhoku Math. Journal* **43**, 249-261.

Sturmfels, B. (1993a). *Algorithms in Invariant Theory*, Springer Verlag, Vienna.

Sturmfels, B. (1993b). Sparse elimination theory, in *Computational Algebraic Geometry and Commutative Algebra* [D. Eisenbud and L. Robbiano, eds.], Proceedings Cortona (June 1991), Cambridge University Press, pp. 264–298.

Sturmfels, B. (1995). On vector partition functions, *J. of Combinatorial Theory, Ser. A*, **72**, 302–309.

Sturmfels, B., Thomas, R. (1994). Variation of cost functions in integer programming, Technical Report, School of Operations Research, Cornell University.

Sturmfels, B., Weismantel, R., Ziegler, G. (1995). Gröbner bases of lattices, corner polyhedra, and integer programming, *Beiträge zur Algebra und Geometrie* , **36**, 281–298.

Tayur, S.R., and Thomas, R., and Natraj, N.R. (1995). An algebraic geometry algorithm for scheduling in the presence of setups and correlated demands, *Mathematical Programming*, **69**, 369–401.

Thomas, R. (1994). A geometric Buchberger algorithm for integer programming, *Mathematics of Operations Research*, to appear.

Thomas, R., Weismantel, R. (1995). Truncated Gröbner bases for integer programming, Preprint SC 95-09, ZIB Berlin.

Trung, N. V., Hoa, L. T. (1986). Affine semigroups and Cohen-Macaulay rings generated by monomials, *Transactions Amer. Math. Soc.*, **298**, 145–167.

Urbaniak, R., Weismantel, R., Ziegler, G. (1994). A variant of Buchberger's algorithm for integer programming. Preprint SC 94-29, ZIB Berlin.

Vlach, M. (1986). Conditions for the existence of solutions of the three-dimensional planar transportation problem, *Discrete Applied Mathematics*, **13**, 61–78.

Wagner, D.G. (1996). Singularities of toric varieties associated with finite distributive lattices, *Journal of Algebraic Combinatorics*, **5**, 149–165.

Weispfenning, V. (1987). Constructing universal Gröbner bases, in Proceedings AAEEC 5, Menorca 1987, *Springer Lecture Notes in Computer Science* **356**, pp. 408–417.

White, N. (1977). The basis monomial ring of a matroid, *Advances in Math.*, **24**, 292–279.

Ziegler, G. (1995). *Lectures on Polytopes*, Graduate Texts in Math., Springer, New York.

Index